The GIS Weasel User's Manual

By Roland J. Viger and George H. Leavesley

Techniques and Methods 6–B4

U.S. Department of the Interior
U.S. Geological Survey

U.S. Department of the Interior
DIRK KEMPTHORNE, Secretary

U.S. Geological Survey
Mark D. Myers, Director

U.S. Geological Survey, Reston, Virginia: 2007

For product and ordering information:
World Wide Web: http://www.usgs.gov/pubprod
Telephone: 1-888-ASK-USGS

For more information on the USGS--the Federal source for science about the Earth, its natural and living resources, natural hazards, and the environment:
World Wide Web: http://www.usgs.gov
Telephone: 1-888-ASK-USGS

Suggested citation:
Viger, R.J., and Leavesley, G.H., 2007, The GIS Weasel User's Manual: U.S. Geological Survey Techniques and Methods, book 6, chap. B4, 201 p; http://pubs.usgs.gov/tm/2007/06B04/

Contents

Figures

Tables

The GIS Weasel User's Manual

By Roland J. Viger and George H. Leavesley

Introduction

The GIS Weasel was designed to aid in the preparation of spatial information for input to lumped and distributed parameter hydrologic or other environmental models. The GIS Weasel provides geographic information system (GIS) tools to help create maps of geographic features relevant to a user's model and to generate parameters from those maps. The operation of the GIS Weasel does not require the user to be a GIS expert, only that the user have an understanding of the spatial information requirements of the environmental simulation model being used. The GIS Weasel software system uses a GIS-based graphical user interface (GUI), the C programming language, and external scripting languages. The software will run on any computing platform where ArcInfo Workstation (version 8.0.2 or later) and the GRID extension are accessible. The user controls the processing of the GIS Weasel by interacting with menus, maps, and tables. The purpose of this document is to describe the operation of the software. This document is not intended to describe the usage of this software in support of any particular environmental simulation model. Such guides are published separately.

Overview

The GIS Weasel is based on the ArcInfo GIS (ESRI, Inc., 2001). The only required input data are raster elevation data, commonly referred to as a Digital Elevation Model (DEM), stored in a single ESRI GRID. The GIS Weasel creates a variety of products, most notably digital spatial data sets, also in the form of ESRI GRIDs, and ASCII parameter files. All interfaces to the GIS Weasel tools are menu and map driven. The user does not need to have any knowledge of the command line operating procedures of ArcInfo Workstation. The system has three major processing phases: *setup, delineation*, and *parameterization*.

The Setup Phase

In the setup phase of the GIS Weasel processing sequence, a variety of topographic surfaces are derived from the user-supplied DEM. The most important of these products is a version of the DEM useful for routing hydrologic flow, a surface of flow direction values, a surface of flow accumulation values, and a map depicting the area of interest (AOI). The details of how these data are derived are described in "The Setup Phase" section. The AOI map serves to spatially limit almost all subsequent outputs and to correspond with the geographic extent of the modeling domain. The AOI is usually a watershed, although it does not have to be topographically delineated or even hydrologically based. The setup phase of a GIS Weasel processing session establishes many of the most commonly needed GIS data sets for delineating geographic features relevant to most environmental simulation models.

The Delineation Phase

Once the setup phase is completed, the GIS Weasel provides tools, through the **Tool Panel**, to delineate maps of different kinds of geographic features within the AOI. Here, the GIS Weasel functions like a toolbox; the user must understand how and when to operate the tools. The user controls what maps or other data are used as input to these tools and can frequently integrate data that were created outside of the GIS Weasel. The user decides exactly how to derive a map of a type of geographic feature, such as a drainage network or watershed. This is where the user must exercise relatively sophisticated understanding of the spatial information requirements of the environmental simulation model and associate these requirements with GIS processing ("geoprocessing") steps or methodologies. The GIS Weasel will not force the association of a geoprocessing methodology with the spatial information requirements of the model onto a user. Although most of the GIS Weasel tools are generic, there are several tools that have been developed in support of specific models, such as the Precipitation Runoff Modeling System (PRMS) (Leavesley and others, 1983) and TOPMODEL (Beven and others, 1995) watershed models. The use of these tools will be documented as separately published guides on the usage of the GIS Weasel in support of specific tasks.

The Parameterization Phase

After the user has created maps of the different kinds of geographic features that depict the geographic domain to be modeled, parameters can be generated about those maps to create a description that the model can use. The GIS Weasel contains a library of hundreds of parameterization routines. The user can apply any of the routines to almost any (ESRI GRID) map, although the relevancy of the output from a routine for a particular simulation model needs to be decided by the user. While the delineation phase of a GIS Weasel processing session is highly interactive, the parameterization phase is highly automated. Once the user specifies the list of parameterization methods and the maps against which the methods will be applied, almost no input is required from the user. The GIS Weasel will produce "parameter maps" showing the derived information. These maps are further processed by the GIS Weasel to produce simple ASCII files containing all the requested parameter information. There is an option in this process to produce an additional file that is specifically formatted for use with the Modular Modeling System (*http://wwwbrr.cr.usgs.gov/mms*). Even if neither of these formats is directly usable by the simulation model, the simple ASCII files have been designed for easy ingestion and reformatting by a variety of external tools such as Microsoft Excel, Microsoft Access, rdb, awk, or perl.

Manual Structure

This manual is intended to serve as a complete reference to all of the functionality of the GIS Weasel. It is likely too detailed to serve as instruction for most beginning users. For a briefer description of the most commonly used functions of the GIS Weasel, the user is referred to the "Sample Usage of the GIS Weasel" section, found later in this document. Versions of example problem sets also are available on the GIS Weasel home page (*http://wwwbrr.cr.usgs.gov/weasel*). The online versions will be kept current with the ongoing evolution of the GIS Weasel software. In addition, the web site includes information relating to the application of the GIS Weasel to specific simulation models, as well as software plug-ins to the GIS Weasel engineered to support these models.

This section is the introduction to the GIS Weasel. The second section, titled "Installation," is a brief discussion of how to install the GIS Weasel and some details about system configuration that a user ought to be aware of prior to running the GIS Weasel. The third section, titled "The Setup Phase," is a detailed description of the GIS theory, menus, and processing steps that comprise the setup phase of the GIS Weasel. Information is described in the same sequence that the user will see during a GIS Weasel processing session. This section is not intended to be used as a "quick-start" guide to the operation of the software. The user should proceed to the "Sample Usage of the GIS Weasel" section for this. The fourth section, "The Delineation Phase," is a detailed description of the GIS theory, menus, and processing steps that comprise the delineation phase of the GIS Weasel. This section exhaustively documents the tools that can be used to interactively query and subdivide the AOI. The fifth section is titled "The Parameterization Phase," and describes how parameters can be derived from the maps created during the delineation phase. As mentioned, the sixth section, titled "Sample Usage of the GIS Weasel," provides an example of an application of the GIS Weasel being used in support of the PRMS (Leavesley and others, 1983). The seventh section, "Advanced Topics," covers a range of detailed discussions intended for more advanced users. Description of the methods used to derive parameters are documented in the "Parameterization Methods" appendix.

Manual Conventions

Where possible, menus and associated graphical components (for example, submenus, lists, buttons, type-in slots) are clearly labeled within the software. The text of this manual will rely on such labels to reference the components. Labels will be written in a **bold typeface**. When referring to subcomponents, the top-level menu is specified and then followed by ">>" and the name of the subcomponent. An example would be the **Quit** button on the **Pan/Zoom** menu, referred to as **Pan/Zoom>>Quit**. The >> convention can be used to specify a series of component nestings. For example **Tool Panel>>Parameterization>>Prepared Models>>Plugins** button refers to the **Plugins** button on the **Prepared Models** menu that is created when the user presses the **Parameterization** button found on the **Tool Panel**.

The main body of text is presented by using a Times font. Text that the user specifies and the names of actual directories or filenames will be written in an *italic typeface*. Text that is generated by the GIS Weasel and written to the MS-DOS Window will be presented in this document in a `Courier Font` and will usually be indented. Text that was generated by the underlying operating system and the names of programs associated with the operating system also are given in `Courier Font`. Technical terms or GIS Weasel related jargon will be presented by using the *italicized* version of the Times font on first usage and also will be defined at that time. The first letter of each word in the term will be capitalized. Jargon also may be used to provide a generic reference to a user-specified value. For instance, the term Input Elevation Grid will be used when referring to the DEM that the user specified at the beginning of the Setup phase. Care has been taken by the authors to differentiate between menu components (for example, **Write Directory and Input Elevation Grid Specification>>Input Elevation Grid**) and the value that a user may specify through a component (for example, Input Elevation Grid) by using these typographic conventions.

Basic GIS Terminology

A brief clarification of the terms associated with GIS in general, the GIS Weasel, and ArcInfo Workstation (the specific GIS software that the GIS Weasel uses) is presented here to help minimize confusion over some of the jargon that is inevitable in this kind of document. There is also a glossary found at the end of the manual. The glossary defines many of the terms used in the "Parameterization Methods" appendix that are related to earth sciences.

A regularly spaced two-dimensional array of points is commonly referred to as a *raster data model*. Within ArcInfo, the raster data model used is called a *GRID*. ESRI GRIDs will always be capitalized in this document to avoid confusing this specific file format with other raster file formats or sets of vector lines arranged into the shapes of square polygons.

Digital Elevation Model or *DEM* is used here as a generic term for a regularly spaced two-dimensional array of points that are attributed with elevation values. This term was originally given to a U.S. Geological Survey-defined ASCII file format for storing elevation data (U.S. Geological Survey, 1986), but has become the accepted term for referencing raster elevation data regardless of the actual file format. DEMs are known by several other names, the most common is Digital Terrain Model (DTM). A Triangulated Irregular Network (TIN) also can be used to represent a surface of elevation, but is not a raster data model. The GIS Weasel does not use TINs. The GIS Weasel uses GRIDs of elevation.

The points that constitute a raster data set are commonly referred to as "cells" and are visualized as boxes. This is because the entire area within the cell is assumed to have the same value as the point around which the box is centered. When the point is being referenced, it is referred to as the "centroid." The distance between adjacent points in a row or column is the cell size of the DEM, and is used as an indicator of the resolution of the data. The distance between a centroid and a diagonally adjacent neighbor is not the resolution or cell size.

As mentioned above, a GRID is specific encoding for a raster data model. A *Lattice*, also an encoding of a raster data model, is similar to a GRID. The salient difference between a GRID and a Lattice is that a Lattice cell is not geo-referenced to a point at the middle of the cell, but to a point that is located at the lower left corner -- the origin -- of the cell. The GIS Weasel uses the GRID data model.

GRIDs can be used to house several kinds of data. The first is a *surface*. The data contained in a GRID surface is likely to be continuously varying. An example of such a surface is a DEM. Every cell within the GRID has the potential to have a unique value. The value is not used as the identity of the cell, but rather as an attribute of the cell. Because the cells are not explicitly identified, the cells are not differentiated or grouped.

The second kind of GRID data is termed in this manual a *zone map*. Within a zone map, cells are grouped and numbered to represent specific entities. Numbers are identifiers in a zone map. All cells having the same numerical identifier are treated as constituting a single behavioral unit, called a zone. A zone can be thought of as raster analog to a feature such as a point, line, or polygon within a vector coverage (described later). All three of these geometries can be easily replicated within a raster zone map. The cells of a zone may or may not be arranged in a single spatially contiguous shape. A single zone identifier may refer to several discrete clusters or islands of cells. Regardless of the spatial arrangement of the cells constituting a zone, the zone is a singular instance; in more object-oriented language, a zone is a discrete object. Generally, the zones in a zone map all depict the same type of geographic feature. For instance, a single zone map will contain only links in a drainage network or roads, but not both. For a further discussion on this, refer to the "Zone Maps Revisited" section of the fifth section on "The Parameterization Phase."

The third kind of data that can be encoded into a GRID is a *categorical surface*. Categorical surfaces are not explicitly identified as a unique type of GRID, rather categorical surfaces are treated by the GIS Weasel as one of the two previously defined types. A common example of this type of data is a GRID of land-cover categories. The number assigned to a cell serves as a code representing the land cover found at that location. Typically the number assigned to a cell is not the identifier of a singular instance or object, but rather is indicative of a characteristic at that location. In this case, the categorical GRID is treated as a surface within the GIS Weasel. An example would be if such a GRID were analyzed to determine the most commonly occurring land cover for each zone in another GRID (of zones) that represents watersheds. On the other hand, if a user's processing of the values in the land-cover GRID treats each value as the identifier of a discrete object, then the land cover functions as a zone map. An example of this would be if the median elevation were derived for each type of land cover in the land-cover GRID. Therefore, in the GIS Weasel, a categorical GRID is treated as a surface by default, but there are user-created situations in which a categorical GRID is treated as a zone map.

A GRID with a limited range of integer values almost always has an associated table, referred to as Value Attribute Table (VAT), that enumerates those values. Zone maps should always have a VAT. Categorical surfaces usually do, unless the value codes used are not integers. By default, a VAT has two columns or fields (also referred to as *items*). The items are VALUE and

COUNT. The VALUE item specifies the numbers associated with one or more cells somewhere within the GRID. COUNT gives the number of cells within the GRID that have the VALUE specified on the same row of the VAT. Having the same VALUE associated with a group of cells is no guarantee that all the cells are arranged in a spatially contiguous shape.

A *cover* or *coverage* is an ArcInfo vector data model. A cover can contain several different types of features, including points, arcs, and polygons. GRIDs and covers are collectively referred to with the umbrella term *geodatasets*. Although a GRID and cover can contain the same thematic information and there are transformations between these data models, these two types of data should not be assumed to be interchangeable. Covers are not usually appropriate for representing maps of information that are continuously varying. The GIS Weasel can display coverages, but generally does not use them for analytical purposes. The GIS Weasel, because of limitations in ArcInfo Workstation, does not support the display or analysis of ESRI Shapefiles, Geodatabases, or Spatial Data Engine services (ESRI Inc., 2001).

Zone Map is jargon designed to alleviate the user from thinking about the differences between raster (that is, GRID) and vector (that is, coverage) representation of geographic features, by providing a term that is neutral with respect to how the features are encoded by the GIS Weasel. In reality, all of the zone maps are encoded in a GRID format. *Zone map* also has been selected to avoid using language specific to any one simulation model. *Feature Map* might have been an equally acceptable alternative to zone map, as well.

The GIS Weasel reads data from and writes data to an ArcInfo *workspace*. Workspaces are standard directories that have been modified for use with ArcInfo Workstation. The user should refrain from manipulating a workspace or the contents (for example, GRIDs or covers) with operating system commands such as `cp` or cutting-and-pasting in `Windows Explorer`, as these commands have no knowledge of the ArcInfo specific details of workspaces. Data contained in workspaces that have been manipulated by operating system commands can easily be corrupted. Further, ArcInfo geodatasets are directories that have been augmented for use with ArcInfo. Therefore, these data sets should not be manipulated with operating system commands. ArcInfo commands should be used when manipulating workspaces or geodatasets. For further information, refer to the ArcInfo Workstation online help (called *ArcDoc* here). In particular, the article can be found in the online help by navigating the following sequence of interactions with it: **Contents>>Arc/Info Concepts>>Arc/Info Data Management**, **Contents>>Arc/Info Concepts>>The Arc/Info Workspace**.

The term *coordinate system* refers not only to a projection and the associated parameters, but also to the datum, spheroid, lateral (XY) and vertical (Z) units. A detailed discussion of coordinate systems is well beyond the scope of this manual. For further detail, the user is referred to **ArcDoc>>Contents>>Arc/Info Concepts>>Map Projections** and Snyder (1987).

Installation

The GIS Weasel software package can be downloaded from the GIS Weasel web page (*http://wwwbrr.cr.usgs.gov/weasel*). The GIS Weasel has been designed to run using version 8.0.2 or higher of ArcInfo Workstation and the GRID extension. ArcInfo Desktop, which is the Windows-only portion of the ArcGIS package installed separately from ArcInfo Workstation, is not required to run the GIS Weasel. The ArcView 3.x software is not used by the GIS Weasel. It should be noted that although ArcInfo Workstation is available for a variety of Unix platforms, it is not currently (2006) available for Linux. The GIS Weasel also uses a small number of C language executables and a variety of other scripts. These are provided with the GIS Weasel software and require no special features of the underlying computing platform. On Unix platforms, the C programs are compiled the first time the GIS Weasel is started. On Windows platforms, the C programs are delivered with the GIS Weasel in a pre-compiled form. Alternatives for hardware can be found at *http://gio.usgs.gov/egis/arc9/install/index.html#plan_and_configure*.

The software is packaged for distribution in a compressed archival format. The same package is used for all computing platforms (for example, Windows XP, and Solaris). No root or administrator privileges are needed to install the GIS Weasel. The installer needs only write-access to the location where the software is being placed; this location can be a user's home directory or shared disk space. Unix users can unpack the distribution file by decompressing it with `unzip`. On a Windows platform, standard packages such as `WinZip` (http://*www.winzip.com*), will handle this operation. The GIS Weasel distribution package will yield a *weasel* directory wherever the user unpacks it. Opening the package within a directory named *weasel* is not necessary; this will result in the software being installed in *.../weasel/weasel*. Although this will not cause problems for the software, this can be confusing to subsequent users and is discouraged.

Once installation (described later) is complete, users are strongly encouraged to work through the "Sample Usage of the GIS Weasel" section. This section serves as a general introduction to the GIS Weasel. It can be accessed through the GIS Weasel home page. The GIS Weasel installation procedures will be described separately for Unix and Windows platforms.

Unix

Download the GIS Weasel software distribution from the GIS Weasel home page to the location where the software will be unpacked. Move, (that is, *cd*) to the location where the GIS Weasel distribution packaged has been placed. Type:

unzip weasel-*.zip

The asterisk ("*") represents the version number of the GIS Weasel distribution package. The ownership and privileges associated with the software will be based on the user-id that unpacked the software. **The user who installs the software should be sure to start the GIS Weasel once.** The first time that the GIS Weasel is run, it attempts to compile the C programs. Once the initial menus appear, the user may press the **Quit** button to immediately exit the GIS Weasel. If the GIS Weasel is not run by the installer, then the C code will not be compiled. In this case, when the GIS Weasel is subsequently run by a different user, the new user will typically not be able to compile the C code (it is owned by the user-id that unpacked the software). Therefore the installer should start the GIS Weasel once to ensure that the C programs are compiled. This is especially important if a system administrator is centrally installing the GIS Weasel for users to access.

At this point, the installation is largely complete. The start-up script is named *weasel* and is found within the *weasel* directory. The installer may choose to add the *weasel* directory to the PATH environment variable for themselves or their users, or to explicitly specify the pathname to the weasel start-up script (*.../weasel/weasel*) each time they start the GIS Weasel software. The weasel start-up script should not be moved from the default location. Problems during start-up of the GIS Weasel usually fall into one of three categories: (1) Compilation of C Programs, (2) Invoking Arc, and (3) System Configuration.

Compilation of C Programs

On Unix platforms, the GIS Weasel C programs are compiled with the user's default C compiler, referred to as *cc*. If the user prefers a different C compiler, the alternate compiler name may be substituted for the original value of "*cc*" in the GIS Weasel Makefiles, which exist at the paths:

.../weasel/src_c/Makefile	(line 17)
.../weasel/src_c/weasel_mms_dmi/Makefile	(line 13)

The line numbers specified to the right of the pathnames given above indicate which line of the corresponding file will contain the C compiler specification. Once the C code is compiled, any user should be able to run the C language executables. If the installer cannot start the GIS Weasel due to compilation problems, it should be confirmed that their environment has access to the standard components for compiling a C program, such as a C compiler, and the `make` and `link` programs.

Invoking Arc

The GIS Weasel start-up script is named `.../weasel/weasel` for Unix users and `...\weasel\weasel.bat` for Windows users, where "…" refers to the location where the GIS Weasel has been installed. Both of these scripts assume that the keyword or command used to invoke ArcInfo Workstation is `arc`. If this is not the case, then the GIS Weasel start-up script will need to be modified. On the 8th line of the weasel start-up script, the *ARC_COMMAND* is set to *arc*. The user should modify the word *arc* to reflect the word used to invoke ArcInfo Workstation on their computing platform. The PATH may need to be augmented to be able to find this command, or included in the *ARC_COMMAND* specification. If the user has questions about this, they should consult with their system administrator.

System Configuration

The GIS Weasel assumes several things about the PATH variable associated with the user's environment, including that the location of the C compiler and the `arc` executable or start-up script can be found within it. The user should consult with the system administrator if these assumptions are not met.

The GIS Weasel assumes that the `DISPLAY` environment variable has been set appropriately. If the user is sitting at the Unix computer whose central processing unit is running the GIS Weasel, then the `DISPLAY` variable is usually not a problem. Alternatively, a user may be remotely accessing a Unix server. In this case, the user or their environment settings must tell the Unix server where to send the GIS Weasel graphical displays. In addition, the user's computer must accept the graphics that the Unix server is sending. This is usually easily resolved. Again, the user should consult with their system administrator.

At several points, the GIS Weasel will make calls to the operating system, known as *system calls*. This allows the GIS Weasel to make use of the operating system, in addition to ArcInfo. The calls are executed within a newly created shell. This shell will be destroyed as soon the task is completed. The user's (or operating system default if none exists) environment settings will be in effect during the existence of the temporary shell. Unusual default shell settings can occasionally cause problems for the GIS Weasel and ArcInfo.

In general, settings associated with the creation of a new shell should append new information to pre-existing environment variables; these settings should *never* re-initialize environment variables. Having a shell environment based on re-initialized environment variables causes augmentations that ArcInfo (or another program) makes to the environment variables to be erased. When the program makes system calls to exploit externally available resources, these augmentations to the environment are often used to find these resources. If the augmentations are no longer in effect (because the environment has been re-initialized), the resources are not found and the system call can fail. Here is an example of an error message that resulted from an environment variable that re-initialized:

ld.so.1: /data2/esri/arcinfo/arcexe80/bin/generate: fatal: libaiaml.so: open failed: No such file or directory

Under Unix, ArcInfo system calls use the C shell. The C shell is configured according to the file, *.cshrc*, which is usually found in the user's home directory. If the user has no *.cshrc* file in their home directory, then a system default *.cshrc* file is used. *.cshrc* lines like the ones shown below have all caused problems for the execution of the GIS Weasel.

if ($?prompt == 0) exit

cd

alias cd `cd {};\'`pwd`:'

PATH=/usr/bin

Aliases of basic operating system commands, start-up commands that change the current directory setting, and re-initialization of environment variables should be avoided.

Windows

Users should download the GIS Weasel software distribution package to the location where it will be installed. The distribution should be then unpacked. Assuming `WinZip` has already been installed on the computing platform, this can be most easily done by right-clicking on the GIS Weasel distribution package within Windows Explorer and selecting the **WinZip>>Extract To Here** option. Press the **Use Evaluation Version** button. Press the **Yes** button in response to the question asking whether `WinZip` should decompress the file to a temporary folder and open it. Other archival software also can be used to carry out this task.

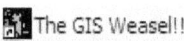

Figure 1.   The **GIS Weasel!!** shortcut.

Windows users have a few more steps beyond this to configure the shortcut that is used to start the GIS Weasel. The user should first locate the shortcut (fig. 1), in the ...\weasel directory using `Windows Explorer`. Once there, right-click on the shortcut and select the **Properties** option from the pop-up menu (fig. 2). The menus may appear slightly differently for different computing platforms (figure 2 shows menus found on a Windows XP computer).

The resultant property sheet is shown in figure 3. There are three properties to adjust: (1) the **Target** (that is, the pathname of the program that the shortcut points to), (2) the directory where the GIS Weasel should **Start In**, and (3) the **Icon**. The **Target** should be modified to point to the actual location where the start-up program, ...\weasel\weasel.bat, is found. The **Start In** directory can point to any directory where the user has write-access. Note for system administrators who have centrally installed the GIS Weasel software, that using a **Start In** location that is local to all machines is needed if network performance is an issue. This also may be important if network drives have not been consistently mounted to the same drive letter. The **Icon** is specified by pressing the **Change Icon** button, and then the **Browse** button on the **Change Icon** menu. On the sub-menu, Look in ...\weasels\src_aml and specify the **File name** of *weasel.bmp*.

Once this is done, press the **Open** button, then the **OK** button, then the **OK** button again (menus will be disappearing, cascading upwards in the menu presentation sequence). At this point **The GIS Weasel!!** shortcut has been configured and the GIS Weasel can be run by double-clicking on it. That will bring up the menu shown in figure 4.

If the GIS Weasel fails to start, the most likely reason is a problem with accessing ArcInfo. A quick test to verify the availability of ArcInfo can be performed as follows:

- Press the **Start** button, at the lower left of the monitor
- Select the **Run** option
- Type *cmd*
- In the **cmd.exe** window, type *arc*

If ArcInfo Workstation fails to start, the user should consult with their system administrator about how to access ArcInfo. The final step that the installer may want to do is to copy the GIS Weasel shortcut to their desktop, although this is not required.

Getting Technical Support

While no guarantee of GIS Weasel performance or support is made, the maintainers of the GIS Weasel do provide support on a case-by-case basis, depending on their ongoing workloads. If the GIS Weasel performs erratically or has an error, then the user is encouraged to email the maintain-

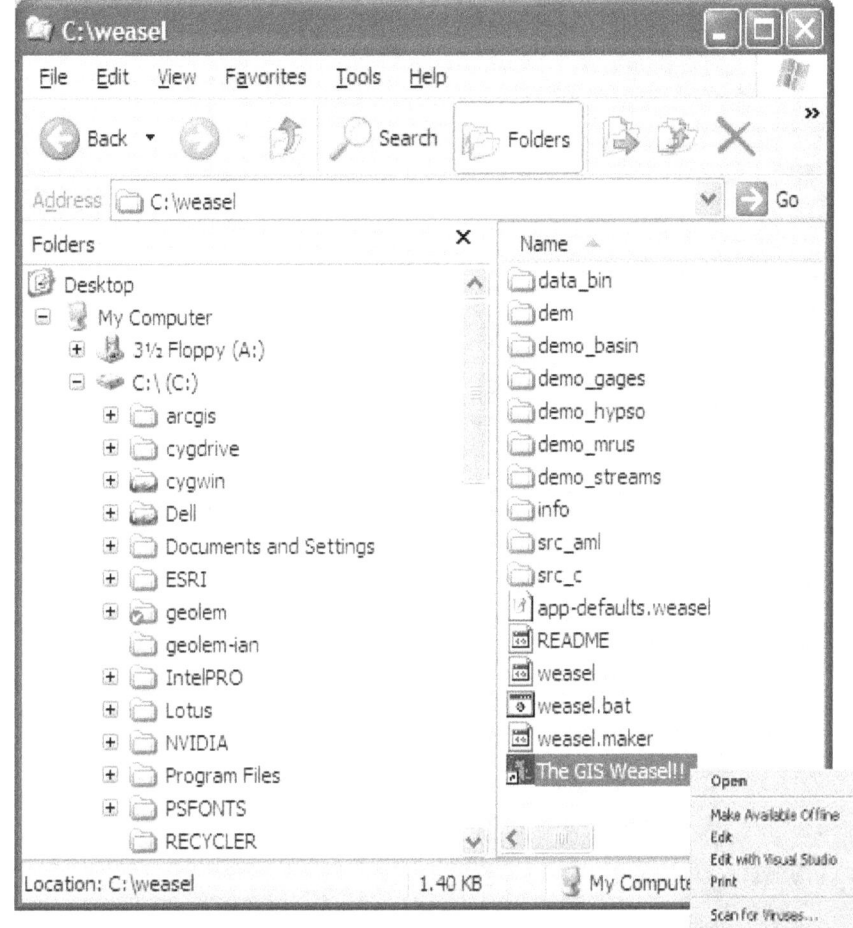

Figure 2.   Accessing the properties of the **GIS Weasel!!** shortcut.

ers of the GIS Weasel with a short description of the last interaction that the GIS Weasel GUI and the problem encountered. A compressed copy of the file, *test.wat*, which is found within the **Write Directory** (described in the next section), should be sent with the email. This file is usually referred to as the *watch file*. Please ensure that the watch file is compressed. It can become quite large and inconvenient for email. Please do not send GIS data. Email can be sent by clicking on the **Your Comments/Bug Reports** link on GIS Weasel home page.

Prior to sending a request for support, the user is strongly encouraged to restart the GIS Weasel and confirm that the problem is recurring and not a one-time anomaly. The user also should carefully validate all of the input data (for example, formats, values, and projections), use of networked drives, write accessibility, or any other factors that are beyond the scope of the GIS Weasel.

Common Problems

This section introduces two of the most common run-time problems associated with the GIS Weasel. The main language that the GIS Weasel uses is the ESRI Arc Macro Language (AML). If the duration of a GIS Weasel session is quite long, then the AML interpreter can run out of memory and cause the GIS Weasel to fail in unpredictable ways. This is usually manifested (and seen in the *test.wat* file) when the GIS Weasel tries to use a variable, but instead of getting the value that has been associated with the variable, a null value is returned. This frequently results in an error message stating that an incorrect number of arguments have been specified (that is, because one of them is missing). This cannot be resolved by the maintainers of the GIS Weasel as there is no fault in the logic of the AML programs. Restarting the GIS Weasel allows the user to exploit all previously completed work with a relatively fresh AML interpreter. This, unfortunately, is the most common reason for the GIS Weasel to

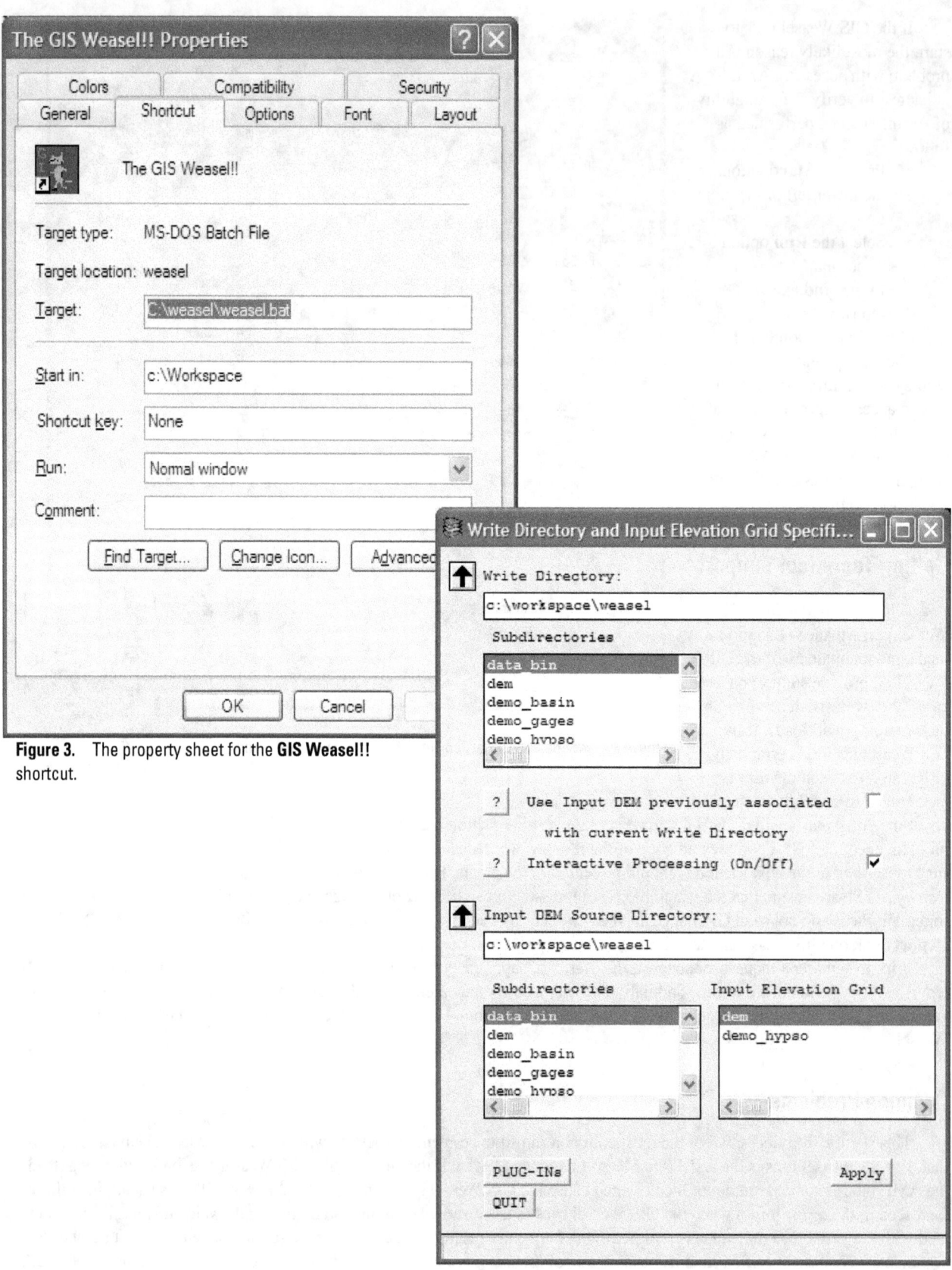

Figure 3.   The property sheet for the **GIS Weasel!!** shortcut.

Figure 4.   The **Write Directory and Input Elevation Grid Specification** menu.

fail. Restarting the GIS Weasel does resolve many of these types of issues. The user should *not* destroy any of the data products made prior to the problem.

Other potential problems for the GIS Weasel relate to pathnames. Extremely long pathnames to the Write Directory (explained in "The Setup Phase" section) can lead to failures during a GIS Weasel processing session. This is largely due to a limitation of the ArcInfo GIS that a command can contain no more than 1,024 characters. In order to avoid this, the GIS Weasel limits the full pathname of the Write Directory to 40 characters or less. This check is a very effective technique for avoiding this problem, but ultimately it does not prevent users from referencing data that has extremely long pathnames.

The Setup Phase

The execution of the GIS Weasel can be considered in three processing phases, labeled here as: (1) setup, (2) delineation, and (3) parameterization. This section describes the purpose and details of the setup phase. This section also introduces some of the GIS Weasel GUI and data-management concepts.

Overview

The purpose of the setup phase is to establish the input to the GIS Weasel and prepare from this a basic set of data products that will be used to derive a variety of zone maps and parameter values. This phase has been automated to the degree possible and usually only takes a few minutes to carry out. The GIS Weasel requires almost no spatial data preparation to actually begin processing. The only needed input is an GRID of elevation data (DEM). Once these data are identified at the beginning of a processing session, the GIS Weasel will carry out a sequence of steps to condition the DEM, create topographic derivatives from it, and allow the user to specify an AOI. The sequence of steps are named as follows:

- Write Directory and Input Elevation Grid Specification
- Setting an Analysis Window
- Filling Pits in the Elevation Model
- Calculating Flow Direction
- Calculating Flow Accumulation
- Delineation of the Area of Interest (AOI)
- Derivation of Slope and Aspect

All output will be written to the Write Directory or automatically created Output Subdirectories. Please consult the "Advanced Topics" section for more information about how the GIS Weasel manages data by using the Ouptut Subdirectories.

Write Directory and Input Elevation Grid Specification

Once the GIS Weasel has been started, the user is presented with a menu labeled **Write Directory and Input Elevation Grid Specification** (fig. 4). Instructions on how to use this menu are presented in figure 5. The *Write Directory* is the location where the GIS Weasel output should be written. The Write Directory specification can be changed by typing into the **Write Directory** slot, by using the **Up Arrow** button to the left of the slot, or by using the **Subdirectories** list below the slot, to navigate to the appropriate location. The default Write Directory is specified by the **Start In** property associated with the **GIS Weasel!!** shortcut (see the

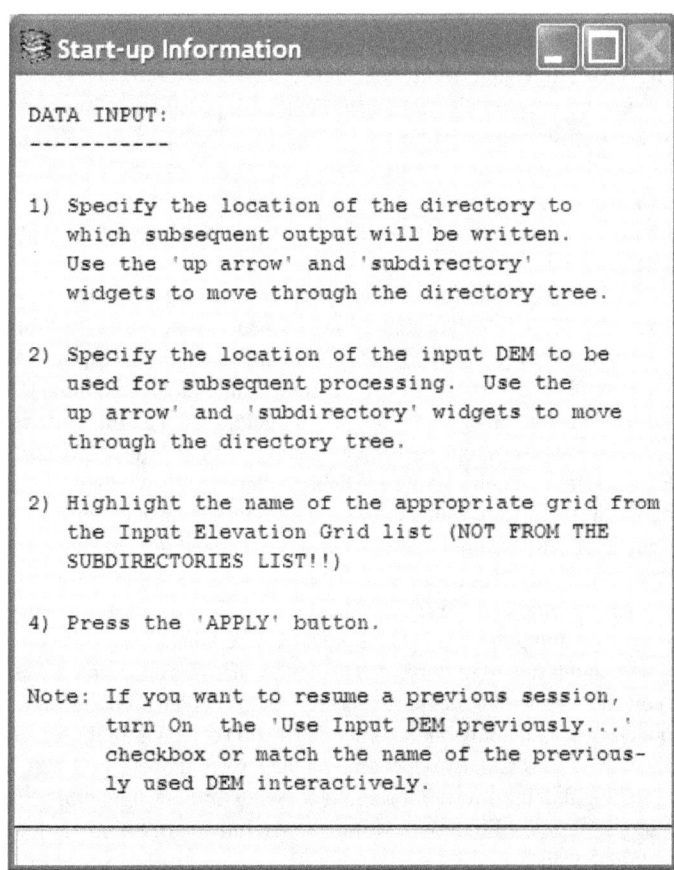

```
Start-up Information                    _  □  ✕

DATA INPUT:
-----------

1) Specify the location of the directory to
   which subsequent output will be written.
   Use the 'up arrow' and 'subdirectory'
   widgets to move through the directory tree.

2) Specify the location of the input DEM to be
   used for subsequent processing.  Use the
   up arrow' and 'subdirectory' widgets to move
   through the directory tree.

2) Highlight the name of the appropriate grid from
   the Input Elevation Grid list (NOT FROM THE
   SUBDIRECTORIES LIST!!)

4) Press the 'APPLY' button.


Note: If you want to resume a previous session,
      turn On  the 'Use Input DEM previously...'
      checkbox or match the name of the previous-
      ly used DEM interactively.
```

Figure 5.   The **Start-up Information** menu.

"Installation" section for details) for Windows users. For Unix users, the default Write Directory is the location from which the GIS Weasel was invoked.

To create a new Write Directory, type the full pathname of the new location into the **Write Directory** slot and press the <enter> key. The user will then be notified that this location does not currently exist and to confirm that the location should be created.

There are several simple guidelines concerning the Write Directory that should be observed by users. The user, of course, must have write access permissions to their Write Directory (and also the **Start In** location). Using pathnames with spaces in them can cause problems and should be avoided. The fully specified Write Directory pathname should be no more than 40 characters long. For Unix users, capital letters should be avoided. All ArcInfo commands are interpreted as lower case letters, and, therefore, capital letters cannot be resolved by ArcInfo (or the GIS Weasel).

The checkbox labeled **Use Input DEM previously associated with current Write Directory** is intended for returning to a previously established Write Directory and the continuation of processing. Turning *On* (that is, "checking") this button will read the metadata associated with the currently specified Write Directory and set the Input Elevation Grid accordingly. The user will be notified if no such metadata record exists and will be forced to choose an elevation GRID before proceeding. The **Question Mark** button to the left will present a contextual help menu on the **Use Input DEM previously associated with current Write Directory** checkbox.

The **Interactive Processing (On/Off)** checkbox indicates whether the user should be asked to change or approve default values in subsequent processing. Leaving this in the *On* position (that is, "checked") is strongly encouraged during the first GIS Weasel processing session within a given Write Directory. This increases the number of times that a user is shown and is asked about the nature of the input data and reduces the unintentional use of potentially inappropriate or incorrect data. The **Question Mark** button to the left of the checkbox will present a contextual help menu.

For users returning to a previously established Write Directory who want to resume where they finished, it can be convenient to turn *Off* the **Interactive Processing (On/Off)** checkbox. This causes all of the previously specified choices and the resultant data products to be automatically reused.

The lower half of the menu is designed to allow a user to specify the location of the GRID of elevation data that should be used. The **Input Elevation Grid** list, at the lower right, shows the names of all recognized GRIDs found in the currently specified Input DEM Source Directory. The Input DEM Source Directory specification can be changed by typing a new location into the **Input DEM Source Directory** slot, or by navigating to a new location by using the **Subdirectories** list below the slot or the **Up Arrow**, to the left of the slot.

GRIDs are actually more than a single file. GRIDs are composed of a set of files and an enclosing directory. This can create confusion for users because the name of the elevation GRID can appear in both the **Input Elevation Grid** list and in the **Subdirectories** list. In figure 4, the name *dem* appears in both places. The user should take care to click on the name of the elevation grid within the **Input Elevation Grid** list. If by accident, the name of the elevation GRID is clicked within the **Subdirectories** list or is specified in the **Input DEM Source Directory** slot, then the user will be reminded that this is not possible and the input will be ignored.

A few brief comments should be made about the characteristics of the **Input Elevation Grid**. It is assumed that the user is familiar with the basic vocabulary and concepts discussed in the "Introduction" section. The GIS Weasel places no real restrictions on the choice of cell size or coordinate system of the **Input Elevation Grid**. Generally, the user is discouraged from using a "geographic" coordinate system (that is, using latitude and longitude to express position), although there are no explicit barriers to doing this. In addition, the vertical units should be the same as the horizontal ones (for example, meters).

The user also should consider the cell size of the **Input Elevation Grid** carefully. The advent of relatively high-resolution data sets (for example, 10-meter DEM resolution) is obviously a great benefit to some kinds of analyses, but can often present logistical issues, consuming large quantities of human and computer processing time, as well as large quantities of disk space. Many applications simply do not need such detailed information to yield results of a satisfactory level of accuracy. The user is encouraged to clearly consider the coarsest cell size that can still satisfy the analytical needs of the problem. If it is apparent that a lower resolution version of the DEM will suffice, the user is strongly encouraged to resample the data to that lower resolution before beginning the GIS Weasel processing session.

A rough rule of thumb when using the GIS Weasel to process elevation data for nonrouted watershed modeling is to select a cell size that will yield approximately 50,000 cells within the watershed. If routing is to be simulated within the watershed model, then something on the order of 100,000 cells within the watershed is appropriate. These numbers are approximate and can change substantially depending on the type of model (time step, level of process resolution), but the user should avoid simply taking the finest resolution and processing millions upon millions of potentially unneeded cells. In addition to these considerations, there are limitations within the ArcInfo Workstation platform, such as the total number of cells or the number and range of unique values allowed within a GRID.

The **PLUG-INs** button allows specialized external software to be used in conjunction with the GIS Weasel. This button is not generally used. It is intended for advanced users and will be discussed in the "Advanced Topics" section. The **QUIT** button will terminate the GIS Weasel processing session. The **Apply** button should be pressed once the Write Directory and the Input Elevation Grid have been specified and the user is ready to begin processing. After the **Apply** button is pressed, a **Map Display** window is presented by the GIS Weasel, along with an accompanying **Pan/Zoom** menu (fig. 6).

The Pan/Zoom menu and the Field of View

The **Pan/Zoom** menu (fig. 6) will be introduced first because this set of tools is basic to exploring the data, and, therefore to responding to the queries presented by the GIS Weasel through interrogation menus during the setup phase. This tool allows the user to control a window within which the GIS Weasel displays data. This window is referred to in this manual as the **Map Display**, although it is labeled **GRID** onscreen. The **Pan/Zoom** menu allows the user to adjust the geographic extent that is shown (also called the *field of view*) and the thematic content of what is seen. The **Pan/Zoom** menu allows users to gain access to the command line and to stop the GIS Weasel software. There are three sections to the **Pan/Zoom** menu: (1) **View Control**, (2) **Theme Control**, and (3) **Miscellaneous**.

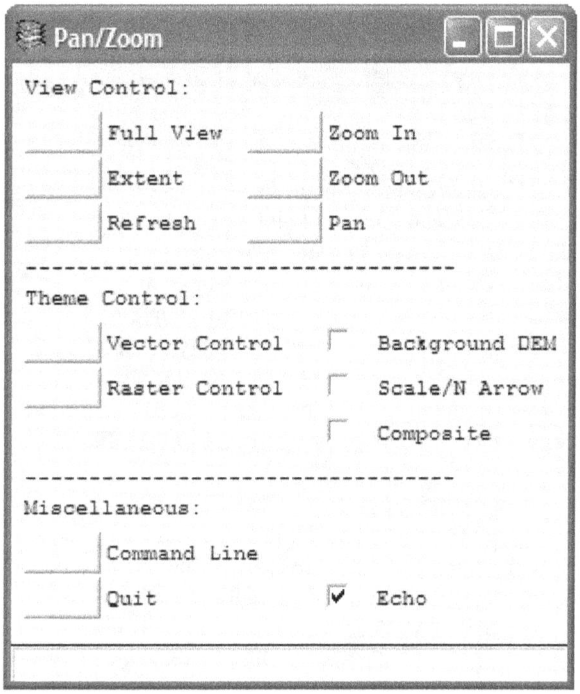

Figure 6.   The Pan/Zoom menu.

Graphical Rendering of GRIDs

The graphical rendering of a GRID in the **Map Display** areas uses either a gray-scale or a default color scheme to symbolize the values. Gray-scale depicts the higher values within the range of data being presented as white, and the lower values as dark. GRIDs that lack VATs are assummed to represent continuous surfaces and are therefore drawn using a gray-scale. GRIDs with VATs are drawn using the default color scheme and each color corresponds to a different value in the VAT. If a GRID has a large number of values, it can be difficult to discriminate between different colors visually.

Pan/Zoom>>View Control

View Control tools function much like the display controls in many other softwares. The **Full View** button resets the field of view to the full extent of the Analysis Window (described in the "Setting the Analysis Window and the Raster Processing Environment" section). This is useful if the user has magnified a portion of the map, but then wants to display the full extent of the data in the Analysis Window. The **Zoom In** button increases the magnification by a factor of 3, centered on a user-selected point within the **Map Display** window. **Zoom Out** decreases the magnification by a factor of .75, centered on a user selected point within the **Map Display** window. **Pan** recenters the display on a user selected point within the **Map Display** window. **Refresh** redraws the current display. **Extent** is the preferred way to increase magnification. With this tool the user must select two points, representing the opposite corners of a new field of view, from the **Map Display** window. This allows the user to control the level of increase in magnification.

Pan/Zoom>>Theme Control

The **Theme Control** portion of the **Pan/Zoom** menu allows the user to change which GRID (that is, *theme*) is shown in the **Map Display** window. There are three less complex display controls available at the right of the **Theme Control** section of the **Pan/Zoom** menu. These are all passive property settings. Changing them will not immediately change the **Map Display** window. These settings are used whenever one of the **View Control** tools is run. The **Background DEM** checkbox, when turned *On* ("checked"), will cause the DEM to be drawn beneath whatever raster data set is currently supposed to be drawn in the **Map Display** window. This can be turned *Off* to increase display speeds. Turning *On* the **Scale/N Arrow** checkbox will cause a scale bar and an arrow indicating north to be drawn on the **Map Display** window. Turning on the **Composite** checkbox will cause the GRID currently associated with the **Map Display** area to be shaded by using the DEM. This tool can be time consuming to run. It is best is left in the *Off* position until actually needed.

There are two types of spatial data that the GIS Weasel can display. These are ArcInfo covers (also called coverages), referred to here as *vector* data, and ArcInfo GRIDs, referred to here as *raster* data. These data types are handled separately, because they have different display characteristics.

Pan/Zoom>>Theme Control>>Vector Control and Display

The **Pan/Zoom>>Vector Control** button invokes the menu shown in figure 7. This menu allows the user to select a single ArcInfo cover (that is, vector data) to superimpose onto the raster data currently shown in the **Map Display** window. This tool can be used repeatedly to add as many covers to the **Map Display** window as the user would like. ESRI shapefiles and geodatabases are not usable by any component of the GIS Weasel.

The **Vector Control** interface works much like **Write Directory and Input Elevation Grid Specification** menu. The **Cover** list shows the ArcInfo covers found in the currently specified Source Directory. This specification can be modified by typing into the **Source Directory** slot, or navigating to a new location by clicking on an entry in the **Subdirectories** list or moving up a level in the directory tree with the **Up Arrow** button. Once a choice has been made from the **Cover** list, the **Symbol** list will be updated to display either line or point symbols depending on the type of data found in the **Cover**. Many more symbols are available than what can be seen within the top few lines of the **Symbol** list and that these may be accessed by using the scroll bar at the right of the list. Once the cover and symbol have been chosen, pressing the **Apply** button will display this information in the **Map Display** window and dismiss the **Vector Control** menu. The **Close** button will dismiss the **Vector Control** menu without adding any new information to the **Map Display** window.

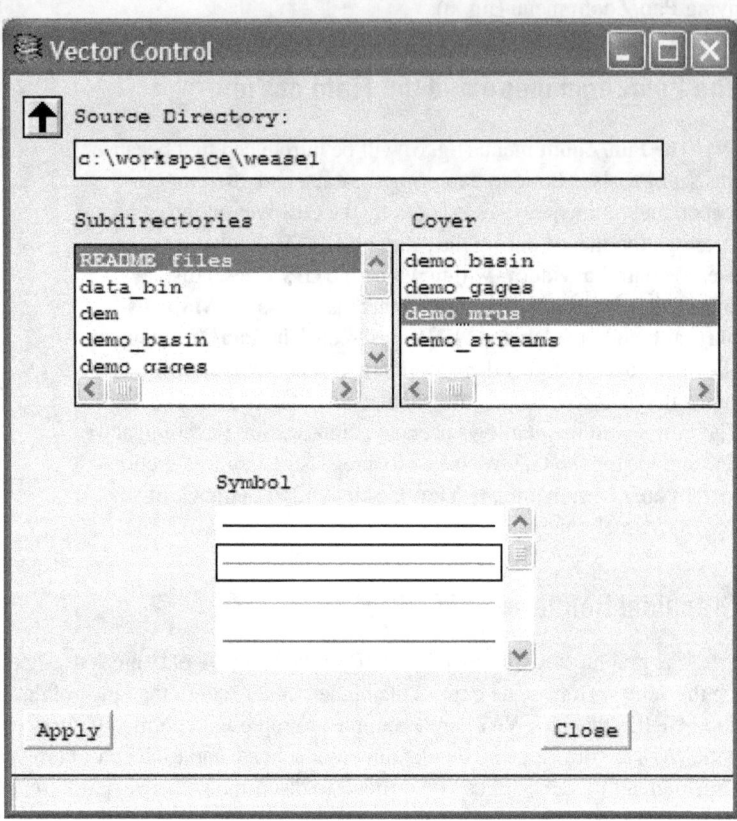

Figure 7. The **Vector Control** menu.

As mentioned, the **Map Display** window updates if the **Apply** button is pressed. In addition, a new menu, labeled the **Vector Display** menu (fig. 8), is presented. If a **Vector Display** menu is already present, then the newly selected cover will be appended to the pre-existing menu.

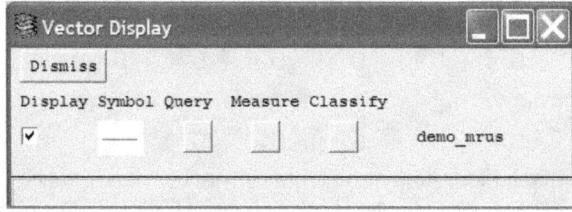

Figure 8. The **Vector Display** menu.

Pan/Zoom>>Theme Control>>Vector Display

Turning *On* ("checking") the **Vector Display>>Display** checkbox indicates that the next time the **Map Display** window is refreshed, the vector data should be displayed. The **Symbol** simply shows the line or point type and color that will be used to present the vector data. The user cannot interact with this feature.

The **Vector Display>>Query** button allows the user to query the features of the vector data by selecting them by mouse clicks in the **Map Display** window. The term *feature* is used here to reference the type of geometry (point, line, or polygon) found in the coverage. Before selecting the individual features, the user specifies which attribute to report during the query (that is, the interactive selection) process. This is done through the the **Vector Display – Query Features** menu (fig. 9), which is produced when the **Vector Display>>Id** button is pressed. The user first selects one of the attributes from the list associated with **Instruction 1**), then presses the **Apply** button. Once this is done, a cross-hair type of cursor will appear over the **Map Display** window. When the user positions the cursor (in the **Map Display** window) over a feature of the vector data set, and clicks with the left mouse button, the value for the attribute of the selected feature is presented in the **Map Display** window and on the **Message Board**, which usually sits at the upper left of the screen (described later in this section). The width of the **Message Board** may need to be increased to see the full message. For Windows users, the last window clicked on tends to move in front of all other windows or menus. This can be inconvenient when working with **Vector Display – Query Features**, because the view of the **Message Board** becomes obscured by the **Map Display** window. Moving the **Message Board** to a location that will not be obscured by the **Map Display** window can avoid this issue. When the user no longer wants to query vector features, the user should click the middle or right mouse button (that is, "nonleft" according to the **Vector Display – Query Features** menu) in the **Map Display** window. If the **Vector Display>>Query** button has been pressed in error, the user can dismiss the **Vector Display – Query Features** menu without any changes by pressing the **Cancel** button.

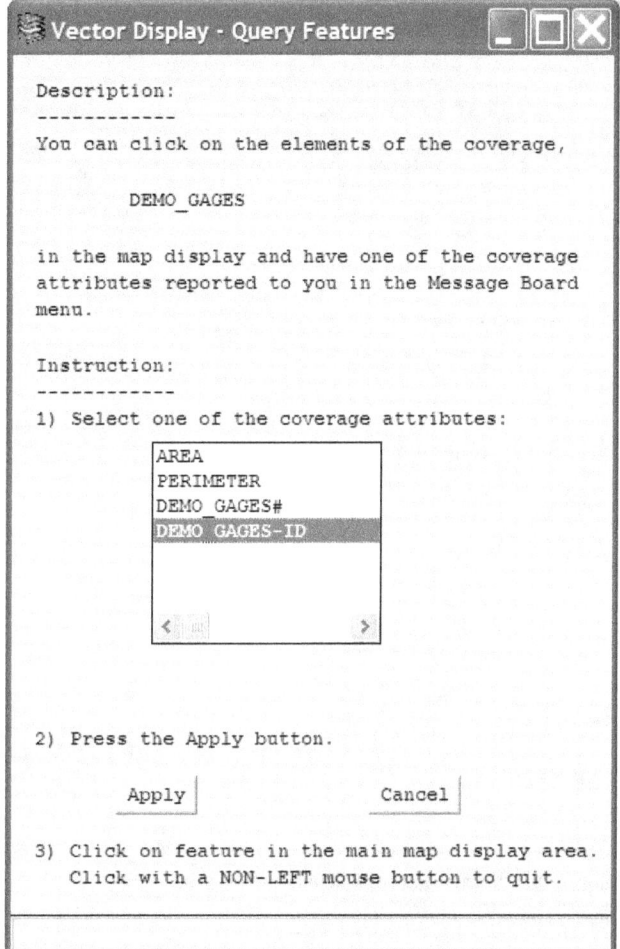

Figure 9.   The **Vector Display – Query Features** menu.

The **Vector Display>>Measure** button is a tool that will calculate either the area or length of user-selected features, depending on the geometry type of the feature. The operation of this tool is almost identical to the **Vector Display>>Id** tool. The **Vector Display>>Measure** tool cannot be used in conjunction with point-type vector information, as this type of data lacks both area and length. If the vector data contains polygons, then the tool will report area. If no polygons are found in the vector data, then length will be reported. Pressing this button will produce the **Vector Display – Feature Measuring** menu (fig. 10). Three steps are required before the user can use this tool. **Step 1** is to choose the units for expressing the area or length. **Step 2** allows the user to specify whether the measure should be rounded off to the nearest integer value before being reported. For example, if the user has selected to have areas of polygons reported in square miles, the user want to get the full decimal expression of the answer (for example, 25.93345) instead of the rounded value (for example, 26). In this case, the user may want to specify *No* for **Step 2** on this menu. Once the first two steps have been completed, the user can press the **Apply** button under **Step 3** to start using the tool. If the user does not want to continue using this tool, press the **Cancel** button.

Once the **Vector Display – Feature Measuring>>Apply** button has been pushed, the user can click on any of the features in the vector data (referred to as the **coverage** in the **Vector Display – Feature Measuring** menu) by using the left mouse button to click on the features in the **Map Display** window. Measurements will be posted to the **Map Display** window and also to the **Message Board**. To stop using this tool, the user should click in the **Map Display** window with either the middle or right (that is, nonleft) mouse button.

The **Vector Display>>Classify** button allows the user to color code (that is, visually *classify*) the features of the vector data according to a feature attribute. This can be a helpful way to look for spatial patterns in the vector information. Figure 11 shows the **Vector Display – Attribute Classification** menu. There are three steps to the application of this tool. **Step 1** allows the user to specify which feature attribute to use. The **Stats** button will produce a description of the number of unique values, the

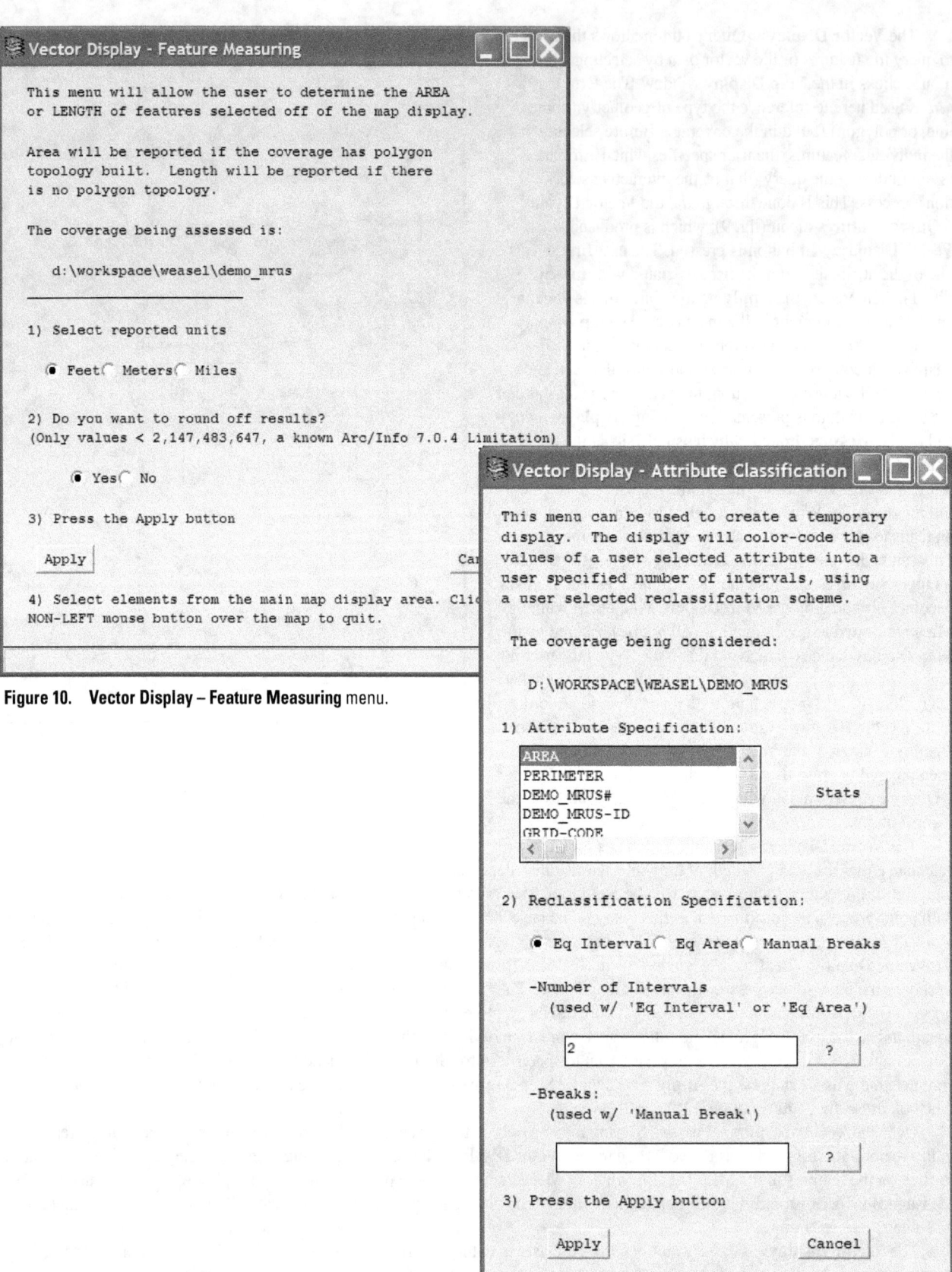

Figure 10.   Vector Display – Feature Measuring menu.

Figure 11.   The Vector Display - Attribute Classification menu.

minimum value, the maximum value, the mean value, and the standard deviation of the current **Attribute Specification**. This is meant to assist the user in decision making about how many intervals are relevant. It is not required that **Stats** tool is used.

Step 2 requires that the user specify the method of classification. '*Eq Interval*' refers to an equal interval classification scheme, where the classification will determine the range of values for the attribute specified in **Step 1** and divide that range into the **Number of Intervals** specified, with each interval having the same size. '*Eq Area*' refers to an equal area classification scheme, where the classification will organize the features in the coverage being considered into the **Number of Intervals** specified, while attempting to ensure that the same number of features will be found in each interval. The total area associated with each interval may not be equal. '*Manual Breaks*' refers to a user defined classification scheme, where the classification will group the attribute values of the features into user-specified bins and color accordingly. The break values, specified in the **Breaks** slot, must be numeric. If more than one break value is specified, the values must be separated by a space (that is, no commas, or tabs).

Pan/Zoom>>Theme Control>>Raster Control

The **Raster Control** button allows the user to select which raster data sets are shown in the **Map Display** window. It should be noted that unlike the display of vector data, in which any number of vector data sets can be displayed on the screen simultaneously, only a single raster data set can be displayed in any given **Map Display** area (although the DEM can be used as a backdrop via the **Pan/Zoom>>Background DEM** checkbox). During the setup phase of the GIS Weasel processing sequence, there is only one area in the **Map Display** window. During later phases, there are four different areas within the **Map Display** window, each showing a different raster data set. The multiple **Map Display** areas will be introduced in the following section.

The **Raster Control** menu (fig. 12) functions in a similar fashion to the several types of data browsers already introduced. The **Input Grid** list shows all the ArcInfo GRIDs found in the currently specified Source Directory. The Source Directory can be modified by typing into the **Source Directory** slot by using the **Up Arrow** button or clicking on an entry in the **Subdirectories** list. When a GRID is selected from the **Input Grid** list, the **Feature Tables** list and the **Items** list are automatically updated. The **Feature Tables** list indicates, perhaps most importantly, whether the GRID has a VAT. Not all GRIDs have a VAT, so this table may not be in the list. If the table does exist, the contents may be viewed by selecting it from the **Feature Tables** list and pressing the **List Table** button found immediately beneath the list.

This **List Table** tool can be useful for learning about the contents of a GRID VAT. The VAT will have at least three items in it. The first is the *Record* number. This is really the row number within the table. By scrolling to the bottom of the list, the user can see the total number of unique groupings of cells there are within the currently specified **Input Grid**. The second is the actual *Value* associated with the cells in the spatial display of the raster data set. This is the identification number for the cells. Although the Value is frequently the same as the Record number, this is by no means a requirement.

The third item is the *Count*. This is the number of cells that share the same Value. Because all cells in a GRID have the same size, the Count can be used to calculate the area that is associated with each Value. To determine the actual area, one needs to multiply the Count by the area of a cell (that is, the cell size squared). The user should remember that unlike vector polygons, cells with the same Value do not have to form a spatially contiguous grouping. A single Value may exist in a large number of isolated cells or groups of cells. Looking at the Count can reveal very small groupings that might not

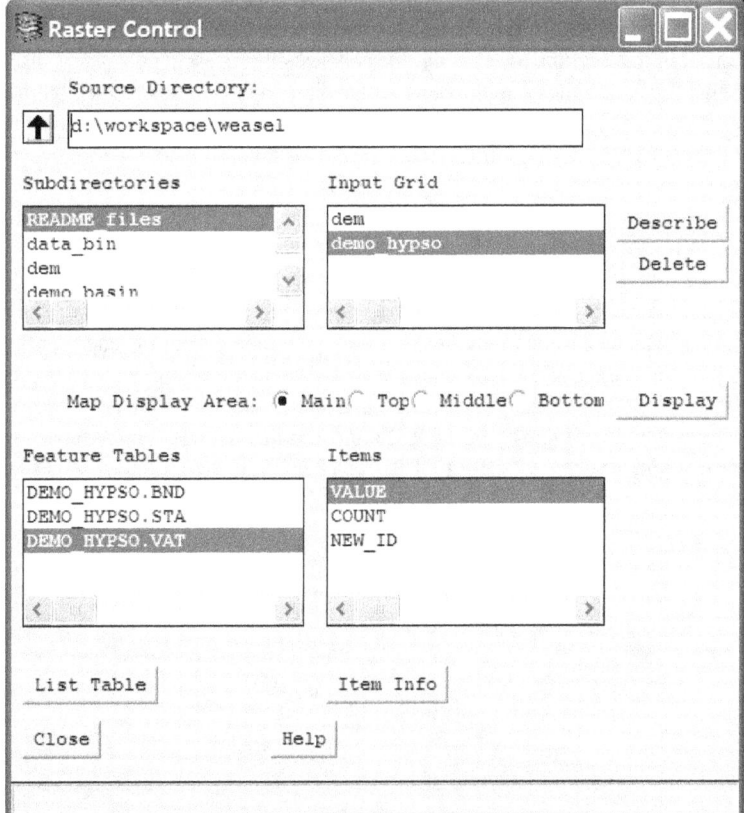

Figure 12.   The **Raster Control** menu.

be apparent from looking at the same GRID in the Map Display window (such small groupings can be missed in a visual inspection of the display).

The **Items** list is generally less important than the other lists. It shows the names of the items (also referred to as *fields* or *attributes*) found in the currently highlighted entry in the **Feature Tables** list. Pressing the **Item Info** button will specify the definition of all the items in the currently specified table. This includes information such as whether the item contains character or numeric information, and the formatting of the fields.

The **Describe** button will present a description of the currently specified **Input Grid**. This description will include information about the spatial coordinate system, the cell size, the range of values found within the GRID, and the physical encoding and organization of the data. The **Delete** button will remove the currently specified **Input Grid** from the hard drive. (Please be careful with this tool. There is no "undo"!) Pressing the **Display** button will cause the currently specified **Input Grid** to be shown in the **Map Display** window. The **Display** button will use the **Map Display Area** setting (*Main, Top, Middle, Bottom*) in the delineation and parameterization phases of the GIS Weasel processing, after the setup phase is completed. The **Help** button will present some more explicit details on the different components of the **Raster Control** menu. The **Close** button will dismiss the **Raster Control** menu.

Pan/Zoom>>Miscellaneous

The final portion of the **Pan/Zoom** menu contains three options. The **Command Line** button is intended for experienced ArcInfo Workstation users who want to access command line functionality of the ArcInfo GIS directly instead of working through the GIS Weasel menus. The **Echo** checkbox controls whether the text of the programs executing within the GIS Weasel DOS-CMD terminal is suppressed or not. In general, the user is encouraged to leave this checkbox in the default *On* position. The text gets recorded in a "watch" file, which is extremely helpful when reporting bugs or seeking support.

Setting the Analysis Window and the Raster Processing Environment

The GIS Weasel is heavily based on raster processing. Whenever the GIS Weasel makes a new GRID, it does so in accordance with a variety of settings. A new GRID will take on the cell size and the coordinate system of the Input Elevation Grid. Another important property is the geographic extent over which data are derived. This also is known as the Analysis Window. Earlier in this section, the effect of the cell size on processing costs was discussed. In fact, the cell size itself does not incur computing cost. The total number of cells within a GRID, determined by the number of times that the cell size divides into the horizontal ("X" and "Y") ranges of the Analysis Window, does incur cost. These quantities are often thought of in terms of the numbers of rows and columns. In the interest of minimizing computing costs, the Analysis Window should be set to the smallest possible area.

After the user has pressed the **Apply** button on the **Write Directory and Input Elevation Grid Specification** menu (fig. 4), the entire geographic extent of the Input Elevation Grid is shown in the **Map Display** window. The user is then asked if they want to adjust the geographic extent over which data are derived by the menu shown in figure 13. If the Input Elevation Grid has already been relatively closely cropped to depict the AOI, then the user can answer *No* to this question. If the extent of the data exceeds the minimum necessary geographic extent needed for the user's application, then the user should answer *Yes*. In this case, a subsequent menu will be presented, asking the user to press an **Okay** button when the field of view (that is, the geographic extent) shown in the map display window is consistent with what the Analysis Window should be. The user can adjust the field of view by using the **Pan/Zoom** menu. (The **Pan/Zoom** menu was described previously in this section.) The **Pan/Zoom>>Vector Control** tool also can be used to superimpose vector layers that may aid in defining the Analysis Window.

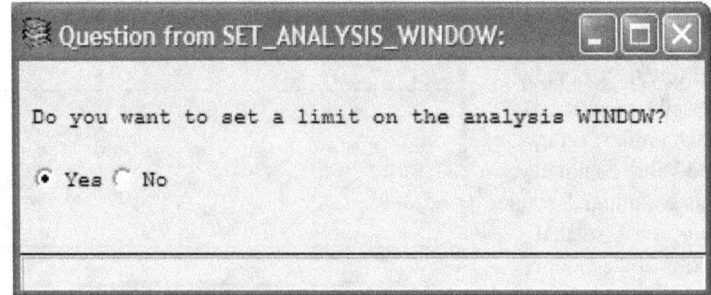

Figure 13. The **Set Analysis Window** menu.

The user is cautioned to avoid the exclusion or truncation of any features or area that will be of interest later in the GIS Weasel processing sequence. The Analysis Window cannot be easily enlarged once the setup phase has been run within a Write Directory. If a mistake is made in setting the Analysis Window, then the GIS Weasel session should be ended (by pressing **Pan/Zoom>>Quit**) and restarted so the Analysis Window can be reset to include the relevant area.

Filling DEM Depressions

Many products derived from the DEM are based on resolving which direction is "most downhill" (sometimes called the *flow direction*, described more later in this section) from each location or cell within the DEM. This is done by comparing the elevation of each DEM cell to the eight neighbors that surround it and determining which of those neighbors, if any, is lower than the cell being analyzed. Of those neighboring cells that are lower, the neighbor to which the slope is the steepest (downhill) is determined. A code indicating the direction toward this neighbor is assigned to the cell in the output GRID.

Frequently, either because of DEM production techniques or subsequent processing (for example, the interpolation used in resampling to a different cell size or projecting the DEM into a new coordinate system), the surface depicted by a DEM may contain spurious crenulations. These crenulations result in low points in the elevation surface that may not exist in the landscape. These low points are sometimes called *sinks*, *pits*, or *depressions*. A pit cell has a lower elevation than the eight immediately adjacent neighbors, and therefore, there is no flow away from the pit. This can cause a variety of higher-level problems for environmental models, as there is no indication within the DEM of how water should move downhill from this point to reach the edge of the geographic region being processed. An *edge* cell is defined here as a cell containing a nonnull data value that is adjacent to a cell containing a null data value. An edge cell can be thought of as existing at the "edge of the world," in that moving away from it will leave the DEM surface.

The ways in which spurious pits effect the creation and description of geographic features depends on the type of analyses that the user chooses to carry out. Although listing these effects is beyond the scope of this manual, the authors would like to point out that pits in the DEM are easily detected by examining the Flow Accumulation surface (described in the "Flow Accumulation" section). Figure 14*A* depicts a Flow Accumulation surface created from an unmodified DEM and figure 14*B* depicts the Flow Accumulation surface created from the filled version of that same DEM. The filled DEM in figure 14*B* has been derived by increasing the elevation values for pit-type cells in the original DEM so that flow can be directed out of these cells. Note in figure 14*A* that there are many relatively short stream-like lines, most of which terminate well before reaching the edge of the DEM, as opposed to the lines of drainage shown in figure 14*B*. The lines of flow accumulation in figure 14*A* lead into pits, from which there is no outward flow.

A

B

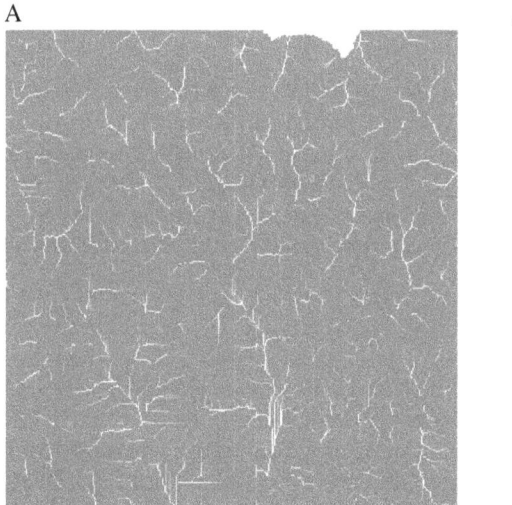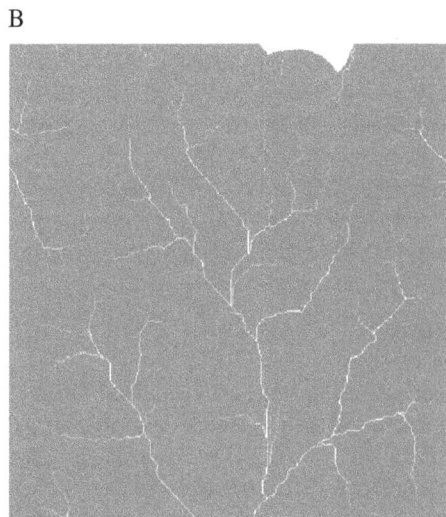

Figure 14.   *A*, The Flow Accumulation surface from an unfilled DEM. *B*, The Flow Accumulation surface from a filled DEM for the same geographic area.

Because topographic derivatives of an unfilled DEM can be problematic, filling the depressions in the DEM is strongly encouraged. Figure 15 shows the menu that is presented to the user, asking whether or not a filled version of the DEM should be created. The technique used to create a filled version of the DEM is an iterative process. On the first pass, all single-celled pits are located and the pit elevations are raised until a flow direction from the pit to one of the immediately adjacent cells can be resolved. In subsequent passes, pits that are larger than a single cell are located and dealt with in a similar fashion. The reader is warned that filling can render large changes within the resultant DEM. More details on the fill algorithm can be found in the **ArcDoc**.

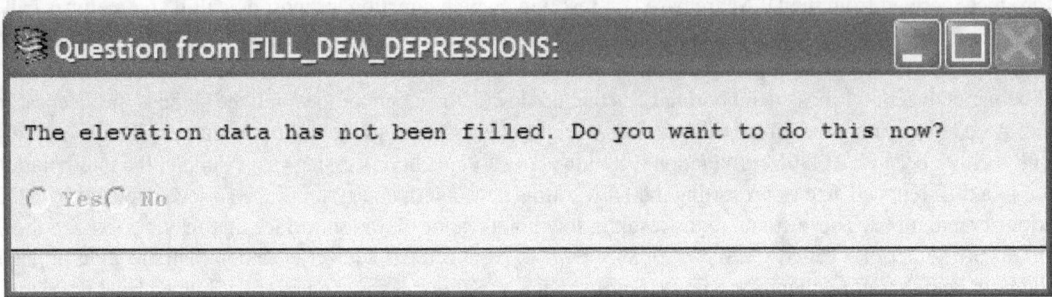

Figure 15. The interrogation menu regarding the filling of DEM depressions.

It should be noted that not all pits are spurious. In the case where a closed drainage area flows to a location on the surface that drains into the ground-water system, as is common in karst (that is, limestone) areas, filling the DEM for this area may have undesirable effects. In the case where there are many pits, some of which are legitimate depressions that must be preserved, the user should process the DEM prior to using it as input to the GIS Weasel. For legitimate pits, users are encouraged to identify the single cell at the bottom of the pit and set its value to null within the Input Elevation Grid. This has the effect of turning that cell into an "edge." In other words, this designates the null cell as an outlet and, therefore, allows water to be routed off the DEM, even if the outlet is in the interior of the DEM and not actually on an edge. The *fill* algorithm still should still be run to remove all the spurious pits that exist in the modified DEM.

The encouragement for filling the DEM depressions should be considered a guideline. Some users may have sophisticated needs for the topographic information that will be extracted from the DEM and will need to assess the effect that filling has on the application. Some may be dealing with areas whose geography simply defies the approach used by the GIS Weasel to resolve flow direction. Areas of extremely locally rough topography, such as those dominated by glacial till that tend to be hummocky or filled with many small ponds, can pose problems. Extremely flat areas, such as the Florida Everglades or other estuarine areas, also can pose difficulties. In some of these cases, the flow direction has been observed to change according to the tides or even the direction of the wind. Unfortunately, neither the GIS Weasel or any GIS can automatically compensate for the diversity of local conditions that might occur. The user is encouraged to simply try a treatment if the effect is uncertain.

If the user elects to use the GIS Weasel to fill the depressions in the DEM, then all subsequent GIS Weasel routines that require an elevation GRID as input will reference this new filled DEM. This newer version will be stored in a grid named *output/surfaces/<my_dem>/filled/grid*, where *<my_dem>* is the name of the Input Elevation Grid and the pathname is relative to the Write Directory. A more detailed description of the Output Subdirectory structure appears in the "Advanced Topics" section.

Flow Direction

As discussed, the elevation of each cell in the DEM can be compared to the eight neighbors surrounding it to determine the direction to the neighbor to which the descent is steepest. This direction is referred to as the *flow direction*. The GIS Weasel requires that a *Flow Direction* GRID be accessible because it is used in so many of subsequent topographic analyses. Therefore, the menu shown in figure 16 will be presented after the user responds to the GIS Weasel query as to whether the DEM should be filled (discussed in the preceding section). Answering *Yes* to the interrogation menu shown in figure 16 will result in Flow Direction surface being derived with the D-8 method (Jenson and Domingue, 1988). Many others exist, although the D-8 method is the most widely used (if only because of incorporation of this method into the ArcInfo GIS). Alternate methods are likely to use different codes to signify flow directions, which will confuse the flow accumulation algorithm used by the GIS Weasel, or any other routines that use the Flow Direction GRID as input. For more information on the details of the D-8 Flow Direction method, look to the **ArcDoc** entry for *flowdirection* (ESRI Inc., 2001).

If the user prefers an alternative type of method be used to generate the Flow Direction GRID used in the GIS Weasel processing session, the user can simply decline the GIS Weasel offer to automatically calculate this data set. If the user declines

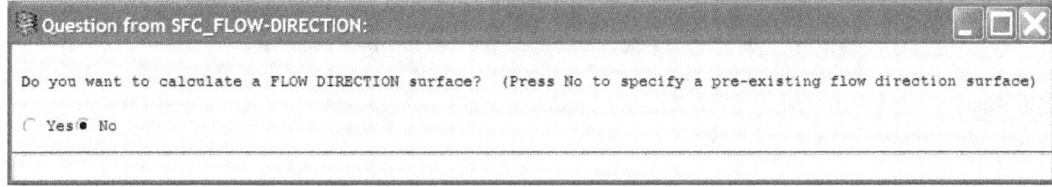

Figure 16.   The interrogation menu regarding the calculation of Flow Direction.

to use the D-8 method, the GIS Weasel will present the **Grid Manager** menu to prompt the user to specify the pathname to the (externally created) Flow Direction GRID to be used. The **Grid Manager** operates exactly as the lower half of the **Write Directory and Input Elevation Grid Specification** menu (fig. 4).

The use of an externally created Flow Direction GRID is not supported, although it is possible to integrate this within a GIS Weasel processing session. If an externally created Flow Direction surface is used, the user also should externally calculate a *Flow Accumulation* GRID (described in the following section) and provide this to the GIS Weasel during the processing session.

If the GIS Weasel calculates the Flow Direction surface, then it will be stored at *output/surfaces/<my_dem>/filled/ flow-direction/grid* or *output/surfaces/<my_dem>/flow-direction/grid*, depending on whether or not the user elected to fill the Input Elevation Grid. *<my_dem>* is the name of the Input Elevation Grid and the pathname is relative to the Write Directory. If the user is using an externally created Flow Direction surface, the location *output/surfaces/<my_dem>/filled/flow-direction/* or *output/surfaces/<my_dem>/flow-direction/* will still exist, but the Flow Direction surface will persist at the original location. Regardless of whether the GIS Weasel has calculated the Flow Direction surface or it has been externally provided, the path- name of the surface will be recorded in a file called *source.weasel* within the output location.

Flow Accumulation

The Flow Accumulation surface, like the Flow Direction surface, is input to many kinds of subsequent topographic analyses and, therefore, the GIS Weasel requires that this GRID exists. The Flow Accumulation surface, derived from the Flow Direction surface, records the number of upslope cells that flow to every cell. This contribution, or number of cells, is often thought of as the contributing area (that is, a watershed) for a cell. The larger the area upslope from a cell, the more water (for instance) will be found in that cell. The farther away from a topographic high point or ridge a cell is, the higher the flow accumulation is likely to be. In places like valley bottoms, the flow accumulation can be quite large because the flow from many cells ultimately con- centrates there. For more information on the details of how the Flow Accumulation method works, please look to the **ArcDoc** entry for *flowaccumulation*.

Once the Flow Direction surface is created or identified by the user, the interrogation menu shown in figure 17 is presented by the GIS Weasel. To have the GIS Weasel calculate the Flow Accumulation surface, press *Yes* on the menu. Pressing *No* will result in the user being queried for the pathname to an externally created Flow Accumulation surface.

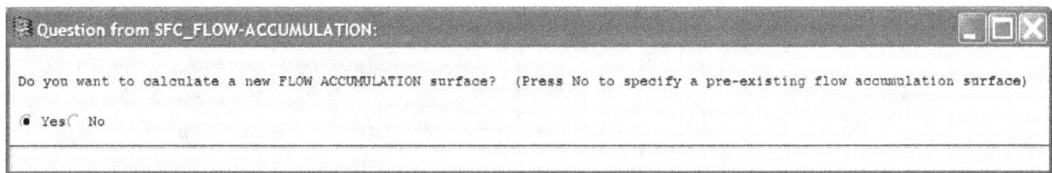

Figure 17.   The interrogation menu regarding the calculation of Flow Accumulation.

One of the most common uses of the Flow Accumulation surface is to find all the cells with a value exceeding some arbi- trarily set threshold and to designate them as being part of a drainage network. There are many other ways of delineating drain- age networks, but this is one of the simplest and most widely used. Figure 14*B* shows a gray-scale display of a Flow Accumula- tion surface, where low values (that is, small contributing areas) are gray and high values (that is, valley bottoms) are white. This color differentiation makes it easy to imagine where the links in a drainage networks should exist. The various considerations associated with picking a specific threshold for the purpose of extracting a drainage network are discussed in the next section.

If the GIS Weasel calculates the Flow Accumulation surface, then it will be stored at *output/surfaces/<my_dem>/filled/ flow-direction/flow-accumulation/grid* or *output/surfaces/<my_dem>/flow-direction/ flow-accumulation/grid*, depending on whether or not the user elected to fill their Input Elevation Grid. *<my_dem>* is the name of the Input Elevation Grid and the pathname is relative to the Write Directory. If the user has used an externally created Flow Accumulation surface, the location *output/surfaces/<my_dem>/filled/flow-direction/flow-accumulation/* or *output/surfaces/<my_dem>/flow-direction/flow-accumulation/* will still exist, but the Flow Accumulation surface will persist at the original location. Regardless of whether the GIS Weasel has calculated the Flow Accumulation surface or it has been externally provided, the pathname of the surface will be recorded in a file called *source.weasel* within the output location.

Area of Interest (AOI) Delineation

Previously, the Analysis Window was set (described in the "Setting the Analysis Window and the Raster Processing Environment" section). This setting bounds the geographic extent over which new GRIDs will be generated. The Analysis Window has the shape of a square or rectangle. Further refinement is normally required to explicitly delineate the spatial boundary for the *Area of Interest (AOI)* of an environmental modeling application. The AOI is solely intended to form the maximum geographic extent of any geographic features that will be described in the user's modeling application. It can be thought of as an irregularly shaped envelope. The AOI can be based on political, hydrological, ecological, or any other geodatasets that are available to the user. Rarely, if ever, does it coincide exactly with the Analysis Window.

Once the Flow Accumulation surface has been identified, GIS Weasel presents the **AOI Delineation** menu (fig. 18). This menu allows the user to define the AOI according to one of two methods from the **AOI Delineation** menu. There is no intent to delineate the specific features within the AOI yet. The delineation of specific features is carried out in subsequent parts of a GIS Weasel processing session (the delineation phase) and will be described in the following section. All the cells within the AOI will be assigned the same arbitrary identifier.

It should be noted that a user may not really need the AOI. It might be the case that no description of the AOI is actually passed to the user's environmental model. Despite this, the GIS Weasel still needs the AOI as a starting point. The AOI is useful for spatially constraining the subsequent delineation of spatial features. For example, if a user intends to eventually delineate streams, this will likely be done by using the Flow Accumulation surface. The values of the Flow Accumulation surface will likely completely fill a square geographic region. The links of a drainage network will be delineated across the entirety of the Flow Accumulation surface. Using the AOI, the GIS Weasel can be instructed to eliminate drainage network links that fall outside of the AOI. An AOI should be made for the benefit of the GIS Weasel processing session, regardless of whether or not the user needs the AOI. The result can be ignored, if desired. See the "Overview on Output of Zone Maps" section in the "Delineation of Features" section for more information.

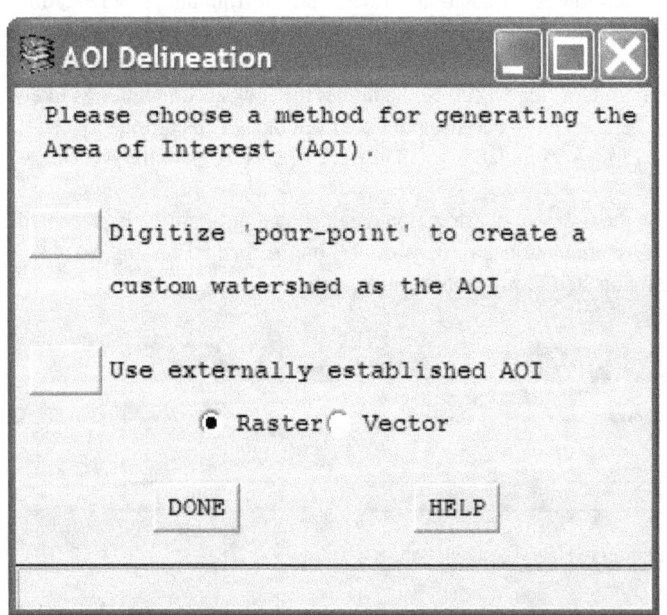

Figure 18. The AOI Delineation menu provides several ways to define an AOI.

There will be two geodatasets produced by the **AOI Delineation** tool, one is the raster version of the AOI and the other a vector version. This redundancy is convenient because the AOI is so fundamental to most model applications. For example, if the user is subsequently examining the DEM but also wants to see the perimeter of the AOI, the vector version of the AOI can be superimposed on the DEM display in the **Map Display** window by using the **Pan/Zoom>>Vector Control** tool. The raster output is named `aoi` and the vector output is `aoi_v`. The "_v" suffix is a stylistic convention that is frequently used within the GIS Weasel to indicate a vector counterpart to a GRID. The user is encouraged to follow this convention for naming vector counterparts to GRIDs. Both the `aoi` and `aoi_v` output will be placed in the Write Directory. The user should not rename these geodatasets, as the GIS Weasel depends on these names to find these data.

Digitize 'pour-point' to create a custom watershed AOI

The first tool on the **AOI Delineation** menu, labeled **Digitize 'pour-point'**…, finds the contributing area above a cell selected by the user from the **Map Display** window. The cell is the contributing area outlet or 'pour-point'. After pressing this button, the **Map Display** window will be redrawn to show the Flow Accumulation surface. The reason for this is that the Flow

Accumulation surface, as described in the previous section, clearly shows where flow is concentrated within a DEM. This has been assumed to be useful to the user, because it is a direct derivative of the Flow Direction surface (and the Flow Direction surface will be used as input to the determination of the contributing area).

The **Pan/Zoom** menu also is accessible during the operation of this tool. It can be used to adjust the field of view within the **Map Display** window so that an individual cell to be used as the 'pour-point' is easy to find (fig. 19). In the example shown, a user is opting to use the cell with the cross-hair symbol on it as the pour-point. This will result in the delineation of an AOI that depicts the contributing area for the tributary that travels from the top of the figure down to the pour-point (and excludes the contributing area for the tributary that travels from the right edge of the figure toward the confluence). The user is free to use the **Pan/Zoom>>Vector Control** tool to superimpose any vector data that might be available in an ArcInfo coverage onto the **Map Display** window, provided that it shares the same projection as the Input Elevation Grid. Stream gages or bridge crossings are examples

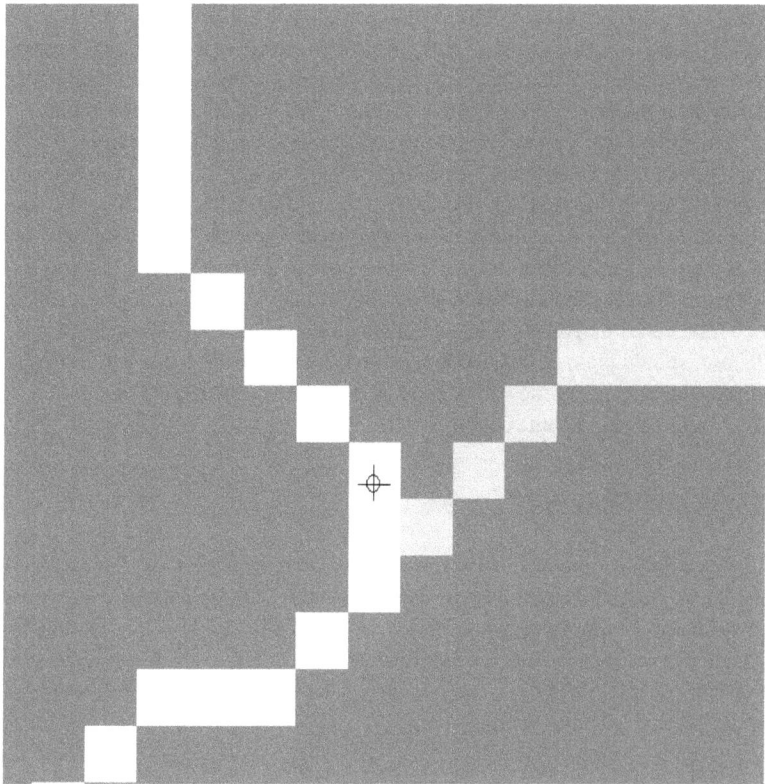

Figure 19. An example of a map extent appropriate for selecting a pour-point.

of the information used to help a user locate the relevant location for a pour-point. In addition, the **Pan/Zoom>>Raster Control** tool can be used to display a GRID other than the Flow Accumulation surface. The user is cautioned that reliably picking a pour-point without some kind of visual cue as to where flow is concentrated can lead to the delineation of a spurious (for example, extremely small) AOI. Once the pour-point has been found and magnified, press the **Okay** button on the **Locate Pour-Point** menu. The mouse cursor will appear as a cross hair within the **Map Display** window. The user should move the cursor over the appropriate cell and click once with the left mouse button.

The **Map Display** window will then zoom out to the full extent of the Analysis Window and show the watershed (that is, upslope area) derived for the selected point. In addition, the area of the watershed will be reported on the **Message Board**. The **Message Board** is usually found at the top left of the screen (hint: these watershed-specific numbers can be compared to the areas reported for other, externally managed features such as stream gages). If the derived area is acceptable, then press the **AOI Delineation>>Done** button to proceed. If the derived area is not acceptable, then press the **AOI Delineation>>Digitize 'pour-point'**... button to try locating the pour-point again. The user also can switch to the second AOI delineation method (explained later).

Using an externally established AOI

The second method for AOI delineation is accessed by the **Use externally established AOI** button. It allows the user to use any ArcInfo GRID or coverage as the basis from which the AOI is defined. The format that the GIS Weasel accepts depends on whether the **Raster** or the **Vector** checkbox (these are mutually exclusive) is turned on at the time the **Use externally established AOI** button is pushed. The user is responsible for specifying a geodataset that shares the same coordinate system and spatial extent as the Input Elevation Grid. If the geodataset is a GRID, there is no need to ensure that it has the same cell size as the Input Elevation Grid.

If the **Raster** checkbox is turned *On* at this time, then a **Grid Manager** will be presented and the user can specify any ArcInfo GRID. If the **Vector** checkbox is turned *On* at this time, then a **Cover Manager** will be presented and the user can specify any ArcInfo cover that has fully built polygons. In other words, if the cover contains only points or lines, then it will not appear in this browser. Sometimes a cover expected to appear in this browser does not. A common reason is that although it contains polygons, the geometry of the cover has been changed since the last time the topological lookup tables associated with that cover were created or updated. The polygons must be consistent with the lookup tables for this tool to detect that the coverage contains polygons. Even if this has been done in the life of a cover, it may need to be repeated if any edits have been made to the coverage since the last time polygons were created. If a coverage is projected to a new coordinate system, polygons also will have to be updated. There are a variety of ways to do this. Within ArcInfo Workstation, the `clean` or `build` commands are available at the `Arc:` prompt. Advanced users may opt to do this by using the **Pan/Zoom>>Command Line** tool.

In either case, if there are multiple areas (that is, zones or polygons) within the geodataset, the user has the option of selecting a subset of areas from the geodataset. If subsetting is declined, then the silhouette of all areas within the geodataset will be used to generate the AOI. The identities of the individual areas will not be maintained within the resultant AOI. The original geodataset will not be modified in any way by this process. If the user opts to use a subset of the areas in the input geodataset, then the user should click on the areas to be used within the **Map Display** window by using the left mouse button. To stop selecting, the user should click with a nonleft (that is, middle or right) mouse button within the **Map Display** window.

The newly derived AOI will be drawn to the **Map Display** window for the user to inspect. If the product is satisfactory, then the user should press the **AOI Delineation>>Done** button to proceed. If the product is not satisfactory, then the user can recreate the AOI with either of the two buttons on the **AOI Delineation** menu until a satisfactory product is created (and then press the **AOI Delineation>>Done** button).

Slope and Aspect

Two GRIDs, one indicating the slope and one the aspect local to each cell, are created by the GIS Weasel based on of the DEM. Because these surfaces are used by many subsequent processes, creation of these data sets is automatic. The GRID of slope will be named *output/surfaces/<my_dem>/filled/slope/grid* or *output/surfaces/<my_dem>/slope/grid*, depending on whether or not the user elected to fill their Input Elevation Grid. *<my_dem>* is the name of the Input Elevation Grid and the pathname is relative to the Write Directory. The GRID of aspect will be stored at a similar pathname, except the last directory is called *aspect* instead of *slope*.

The Delineation Phase

The setup phase will furnish the Write Directory with a geographic extent, a DEM, Flow Direction, Flow Accumulation, and the AOI, as well as slope and aspect surfaces. With these building blocks, the GIS Weasel is prepared to help the user explore the geographic data sets associated with their AOI and carry out the delineation of any number of types of geographic features, ultimately leading to the generation of parameters describing these features to the user's model. Parameter generation is discussed in "The Parameterization Phase" section. Once the setup phase of a GIS Weasel processing session has executed, the software will signal the opening of the delineation phase by adjusting the **Map Display** window, presenting a new version of the **Pan/Zoom** menu, and presenting the **Tool Panel**. Details on all three of these will be provided within this section. The presence of the **Tool Panel** is the most obvious indicator that the setup phase has completed.

In contrast to the setup phase, which consists of a linear sequence of processing steps, the delineation phase is open-ended. How the GIS Weasel should be used after this point is entirely up to the user. Although the GIS Weasel does absolve the user from having expertise with the commands of the ArcInfo GIS, it does not absolve the user from having an understanding of the spatial information used in the environmental simulation model. This understanding should include a sense of how the geographic features, whose behavior will be simulated by the model, are defined. This includes information such as which attributes are used to find and isolate an instance of the feature, if or how the feature relates to others of its kind, if or how it relates to other types of features, how heterogeneous a feature can be with respect to certain attributes, and if there are feature size, shape, or numbering constraints imposed by the environmental model. Although the GIS Weasel development team has bundled software to provide support for specific models, this has only been done for a small number of cases. The GIS Weasel is intended to serve as a generic, stand-alone tool that is useful for a wide variety of environmental simulation model users. The user is referred to the "Plug-ins" section of the "Advanced Topics" section for more information about adding functionality specific to their needs into the GIS Weasel.

For some users, no delineation beyond defining the AOI is needed. In this case, the user might proceed directly to the parameterization phase. Other users might need to delineate a variety of complex zone maps depicting different types of geographic features. For this group, learning about the set of exploration and delineation tools, referred to in this manual as the *delineation phase*, is probably a worthwhile investment of their time. These users are encouraged to consult with the GIS Weasel development team if questions remain after reading this manual and working through the associated online exercises available on the GIS Weasel home page.

Prior to beginning this discussion on delineation, it should be stressed that the application of parameterization tools, in general, is not affected by how a zone map of geographic features was delineated, or even if the delineation was done within the GIS

Weasel (as long as the coordinate systems of all GRIDs are consistent). While the GIS Weasel attempts to present a wide variety of tools for the delineation of geographic features, there are many other software packages available for this task, and it may be more convenient for the user to simply use the GIS Weasel as a parameter generation engine and ignore the feature delineation capabilities. The details on how to generate parameters for externally created GRIDs is addressed in "The Parameterization Phase."

Overview

This section on the delineation phase, as mentioned above, will provide a reference on the purpose and use of tools for (1) the exploration of attributes associated with the AOI, and (2) the delineation of features. The former is relatively brief, composed of only four tools. The latter is extensive, covering almost 30 different tools. All of these tools are accessed from the **Tool Panel**, discussed in the "Tool Panel" section. The **Tool Panel** buttons labeled **Parameterization of features** and **Other Output** will be addressed in subsequent sections ("The Parameterization Phase" and "Advanced Topics," respectively). Also discussed in this section are the changes to the **Map Display** window and **Pan/Zoom** menu that were introduced in the previous section.

This section is not intended to serve as a tutorial for using the GIS Weasel. Users are referred to the "Sample Usage" section for an example of a GIS Weasel processing session based on the sample data distributed with the software. That session shows an example of using the GIS Weasel in support of a specific environmental simulation model.

The Map Display Window and the Pan/Zoom menu

During the previous phase, the **Map Display** window showed only one GRID, largely in the interest of keeping the graphical user interface as simple as possible for the user during a sequence of processing steps from which there is usually very little need for variation from the default experience. During the delineation phase, the user is likely to be examining a variety of GRIDs and simultaneously evaluating their content while making decisions about delineating geographic features. Given this, the **Map Display** window has been enhanced to allow the visualization of four GRIDs at the same time. Figure 20 reveals that there is a larger map display (referred to as the *main* **Map Display** area) and three smaller ones (referred to as the *top, middle,* and *bottom* **Map Display** areas).

Each area within the **Map Display** window has the same capabilities; the main one is simply larger. Each area can display a different GRID and is labeled at the lower left corner with the name of the GRID currently being displayed. If the pathname to the GRID is too long to be displayed, the trailing portion of the pathname will be displayed (see the middle and bottom

Map Display areas in figure 20). Each **Map Display** area can be adjusted by the user to present a different field of view (the field of view also is referred to as the *map extent*). As in the setup phase, any number of vector data (that is, coverage) can be superimposed onto the map. Modifications to each area are effected by using the **Pan/Zoom** menu, described later. As during the setup phase, the **Map Display** window will go blank if resized. The user simply needs to use the **Pan/ Zoom>>Refresh** tool to redraw the content into each area of the **Map Display** window. The user can resize the window to their convenience.

Only the enhancements to the **Pan/Zoom** menu (fig. 21) will be described here. For the original documentation on this menu, the user is referred to the relevant section in the preceding section. The **Pan/ Zoom>>View Control** button operations have already been described,

Figure 20.   The **Map Display** window after the setup phase.extent appropriate for selecting a pour-point.

Pan/Zoom

View Control:

	Full View		Zoom In
	Extent		Zoom Out
	Refresh		Pan
	Switch Main/Sub-Display

- -

Theme Control:

☐ Composite ☐ Raster Control

☐ Scale Bar ☐ Vector Control

☐ DEM backdrop

☐ Number Zone Map

☑ Auto-Display New Zone Map

- -

Miscellaneous:

| | Command Line | | Clean WriteDir |
| | Quit | | AnalysisWindow |

Figure 21. The **Pan/Zoom** menu after the setup phase.

but the operations of the buttons have one additional aspect. Because there are now four areas within the **Map Display** window, the user must identify which of the four areas is being adjusted. This is done by clicking the left mouse button over the desired area, after pressing one of the **Pan/Zoom>>View Control** buttons.

The **Switch Main/Sub-Display** button is a new addition to the **Pan/Zoom** menu since the completion of the setup phase. This switches the GRID that is being displayed in the main **Map Display** area with the GRID being shown in one of the smaller **Map Display** areas. After pressing this button, the user should click the left mouse button over either the top, middle, or bottom **Map Display** areas. This tool is useful if the user has zoomed in the field of view in the main **Map Display** area to allow closer examination of the GRID displayed therein, and then wants to examine the content of a GRID associated with another **Map Display** area for the same location.

Within the **Pan/Zoom>>Theme Control** area, there are two checkboxes that have been added since the completion of the setup phase. These are labeled **Number Zone Map** and **Auto-Display New Zone Map**. The GRID format that is used to store zone maps unfortunately does not support labeling. To enable the numerical labeling of features, the GIS Weasel will create a vector version of the original raster zone map and use this to label the (raster) features with the identification numbers. The **Number Zone Map** checkbox is the tool for controlling this functionality. Turning it *On* (placing a check in it) simply sets a property that will be used the next time any of the **Map Display** areas are redrawn. If this checkbox is *On*, then the GIS Weasel will look for a coverage with the same name as the GRID associated with the **Map Display** area being updated, but with a "*_v*" suffix, within the Write Directory (it doesn't matter where the GRID is actually located). If found, the coverage will be examined for an attribute of the polygons called *GRID-CODE* for use as a label. If no such coverage exists, then the user will be presented with the **Pan/Zoom>>Number Zone Map** menu, shown in figure 22, that will create such a coverage.

This menu will create a vector version of the zone map associated with the **Map Display** area being updated. In the example presented in figure 22, a coverage called *str2kn_v* will be created in the Write Directory. The user is asked to describe the basic geometric type of the data being vectorized. Clicking on an option will result in immediate conversion. Point type features are not supported. If the zone map has single-celled zones that are effectively points, selecting the **Area** option is encouraged. The **Number Zone Map** checkbox is *Off* by default. Generating a vector version of a GRID can sometimes be a very time-consuming task if there are many discrete zones (for example, 5000) in the zone map. If the GRID being displayed was a surface, such as the DEM, the process of trying to number every cell could result in an unwanted termination of the GIS Weasel processing session. Another possible drawback to having this tool constantly active is that it can clutter the graphical display.

If there are more than 300 zones in the zone map, then no labels will be placed. If the area of a zone is less than 100,000 square meters or feet (depending on the coordinate system of the Input Elevation Grid), then it will not be be labeled.

Figure 22. The **Pan/Zoom>>Number Zone Map** menu.

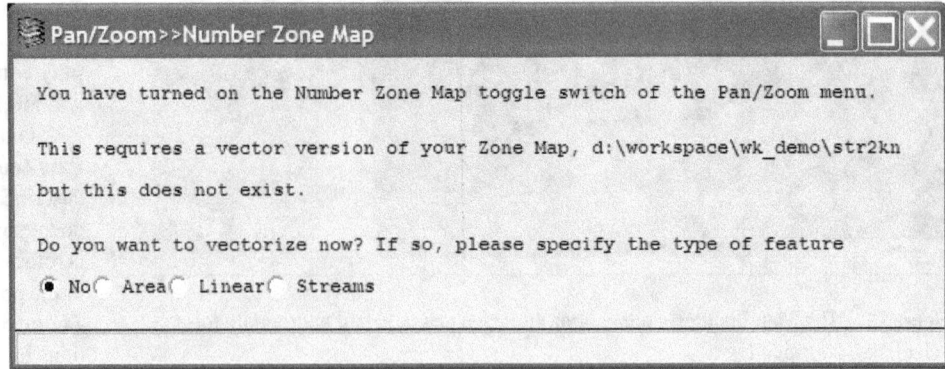

Pan/Zoom>>Number Zone Map

You have turned on the Number Zone Map toggle switch of the Pan/Zoom menu.

This requires a vector version of your Zone Map, d:\workspace\wk_demo\str2kn

but this does not exist.

Do you want to vectorize now? If so, please specify the type of feature

● No ○ Area ○ Linear ○ Streams

If a coverage counterpart exists, but there is no GRID-CODE attribute associated with its polygons, no labeling will occur. If the user removes this coverage (use on of the ArcGIS commands) from the Write Directory, the GIS Weasel will create a new, properly attributed coverage that will allow labeling.

The **Pan/Zoom>>Auto Display New Zone Map** checkbox is *On* by default. This signifies that any time the user generates a new zone map, it is used as the GRID shown in the main **Map Display** area. The generation of a new zone map is described in the "Overview on Output of Zone Maps" section.

The **Pan/Zoom>>Clean WriteDir** button allows the user to safely remove temporary files and grids from the Write Directory, as well as, from the Output Subdirectories. The GIS Weasel automatically creates, names, and stores many intermediate products in the Output Subdirectories for future reuse and for documentation of a user's processing history. It might be the case that the user wants to run a tool "from scratch" and avoid reusing any pre-existing products. This can be done by erasing the pre-existing products with the **Pan/Zoom>>Clean WriteDir** tool. This can be a complicated process requiring familiarity with the GIS Weasel processing logic on the part of the user. The user is referred to the "Advanced Topics" section. The Output Subdirectories also are discussed in detail in the "Advanced Topics" section.

The **Pan/Zoom>>Analysis Window** button allows the user to adjust the Analysis Window that was set during the setup phase. After pressing this button, the user is asked to use the **Pan/Zoom>>View Control** tools to adjust the field of view in the main **Map Display** area. Once the field of view reflects the desired Analysis Window, the user presses the **Okay** button on a menu used to signify that the new Analysis Window has been set. If the extent of the Analysis Window is changed, the extent of the data content found in GRIDs pre-dating this change will not reflect this change. Although GRIDs produced after the Analysis Window is changed will reflect the new geographic extent of the Analysis Window, there might be areas of null values if the input (that is, the GRIDs that predate the enlargement of the Analysis Window) did not have nonnull values throughout the new Analysis Window.

Tool Panel

The Tool Panel (fig. 23) is at the heart of all subsequent processing within the GIS Weasel. There are three major functional groupings of the buttons on this menu. At the top of the menu is a cluster of buttons under the label **EXPLORATION of Attributes**. These tools are not intended to generate new zone maps, but to help the user interact with the raster data and to serve as input to the user's decision-making process about delineating features and generating parameters about those features. The second group of buttons is labeled **DELINEATION of Features**. The buttons in this grouping generate new Zone Maps. Most of the tools represented here are generic enough to be useful to many applications, but a handful have been engineered in support of specific environmental simulation models. The last major functional grouping is the single button labeled **PARAMETERIZATION of Features**. This tool yields a set of new menu components. The details of operation are covered in the "The Parameterization Phase" section. At the bottom of the **Tool Panel** is an additional button, labeled **OTHER Output**. This tool is intended for more advanced users, providing lower-level access to a large variety of the routines underlying the GIS Weasel menus. This tool is described in the "Advanced Topics" section.

At the top of the **Tool Panel**, there is a slot labeled **Current Zone Map**. The Current Zone Map is an important concept. Almost every tool for delineating a new zone map uses a pre-existing zone map as the starting point, even if only to spatially constrain the new product. An example is the delineation of a drainage network. Strictly speaking, only the Flow Accumulation surface is needed to do this. This surface may extend well beyond the area that is defined by the AOI and, therefore, it is likely to yield lines of drainage beyond the AOI. Although the AOI itself does not have anything directly to do with the process of finding the lines of drainage, it is useful for automatically eliminating all the drainage lines that fall beyond the

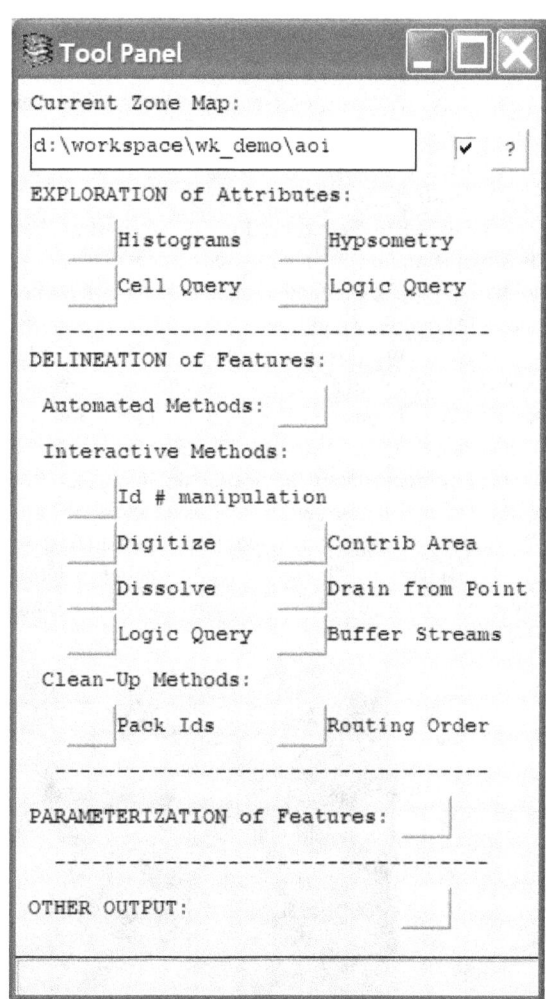

Figure 23.   The **Tool Panel.**

perimeter. To help the GIS Weasel do this, the user should ensure that the **Tool Panel>>Current Zone Map** is pointing to the AOI map before starting the tool for extracting a drainage network.

To specify a new value for the Current Zone Map, the user can type in a new name directly into the **Current Zone Map** slot. Alternatively, clicking the right mouse button in this slot will produce a **Grid Manager**. Whatever name is selected in this browser will be placed in the **Tool Panel>>Current Zone Map** slot. If the checkbox to the right of this slot is turned on, which is the default, then the main **Map Display** area will be redrawn by using this new GRID. It should be noted that the GIS Weasel requires that the Current Zone Map be specified before it will carry out any commands, including **Pan/Zoom>>Quit**.

Exploration of Attributes

Included in the **Tool Panel** is a set of tools that are intended to help the user understand the values of the GRIDs that are shown in one or more of the **Map Display** areas. Most of the following examples will use the DEM or the slope and aspect derivatives, but the tools are not limited to these data sets. The user may apply these tools to any GRID by associating it with any of the **Map Display** areas prior to running the tool. This association is usually made by pointing the **Pan/Zoom>>Raster Control>>Input Grid** to the GRID of interest and pressing the **Pan/Zoom>>Raster Control>>Display** button. The user also can adjust the **Pan/Zoom>>Raster Control>>Map Display Area** property prior to pressing the **Pan/Zoom>>Raster Control>>Display** button. These tools are not intended to make new maps, although the **Tool Panel>>Logic Query** tool will allow the user to store the output in a new GRID.

Histograms

Frequency histograms are useful for providing a visual indication as to the shape of the distribution of values in a GRID, as well as the magnitude and ranges of values and frequencies. The **Tool Panel>>Histograms** tool allows the user to simultaneously create three histograms, one for each of the GRIDs shown in the top, middle, and bottom **Map Display** areas. Rather than use all of the data found throughout the full spatial extent of these layers, the tool can spatially limit which GRID cells are used to construct the histograms on the basis of user-selected zones from the Current Zone Map in the main **Map Display** area. The three histograms will be built by using only cells that are found within the user-selected zones. These zones are treated as a single filter. For each of the GRIDs, the tool will aggregate the data associated with all user-selected zones into one set and construct a single histogram based on this set.

The user can select zones from the main **Map Display** area by clicking in the Current Zone Map with the left mouse button. Clicking the middle or right mouse button will cause the selection process to terminate. If no zones have been selected, then the tool will notify the user and terminate. If zones have been selected, the tool will present a new graphics window to the screen, such as the one shown in figure 24.

There are two possible types of histograms. The left-most histogram (fig. 24) is an example of the floating-point type histogram associated with a GRID that has no value attribute table (VAT). This type of GRID usually represents a continuous surface. Every cell within this type of GRID has the possibility of having a unique value. The most common example of this type of GRID is a DEM. The histogram therefore is generated by using a dynamically created reclassification of the original data into 100 bins. If this reclassification were not done, then frequencies of 1 would be common (because each cell likely has a unique value). The user is referred to the "Introduction">>"Basic GIS Terminology" section for more information on GRIDs and VATs.

The second type of histogram, seen in both of the graphs to the right of the floating-point type histogram (fig. 24), is for GRIDs that have a VAT. Having a VAT implies that there is a limited variety of values found throughout the entire GRID. The values are commonly categories. Land cover GRIDs are common examples of categorical raster data. Because of the limited

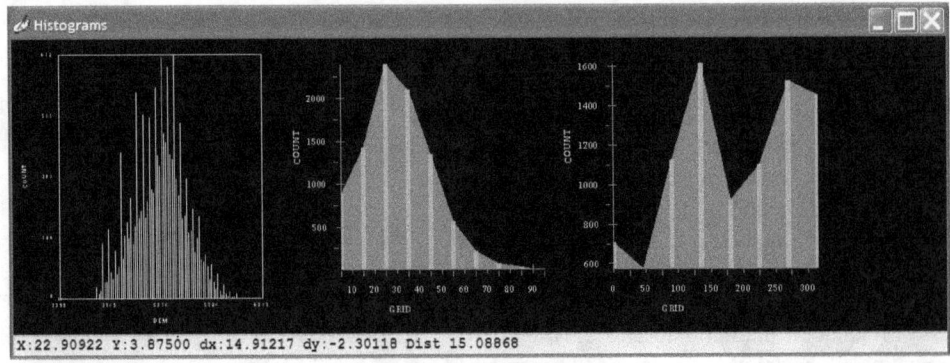

Figure 24.   The graphic output from the **Tool Panel>>Histogram** tool.

number of values, no reclassification is done to support the generation of this kind of histogram. The green columns within the histogram correspond directly to values found in the VAT. The red fill that connects the green columns in the histogram is strictly a graphical device to make the histogram more readable. There is no actual data associated with the red fill.

The GRIDs being represented by the three histograms will always correspond, from left to right, with the top, middle, and bottom **Map Display** areas. There is a label below the x-axis of each histogram that indicates the name of the GRID that was used to create the histogram, although this is not always very useful because many of the GIS Weasel-generated GRIDs are simply called "grid." Remembering the relation between the order of histograms and the **Map Display** areas is the preferred way to understand what data are being displayed in a histogram.

After the histograms have been presented to the screen, a new prompt asks if the user would like to select a new set of zones for generating a new set of histograms. If so, then the **Histograms** window will be redrawn to show the new histograms. If not, the tool will terminate. In this latter case, the **Histograms** window will remain. To dismiss this window, click on the **X** button at the top right corner of the window. Resizing the **Histograms** window will cause the graphics to disappear from the window.

Cell Query

The graphical rendering of the GRIDs in the **Map Display** areas uses either a gray scale or a default color scheme to symbolize the values. The gray scale depicts the higher values within the range of data being presented as white, and the lower values as dark. The gray scale is automatically applied to GRIDs that lack a VAT, assuming that these are continuous surfaces. Where a GRID does have a VAT, the color scheme is applied using the assumption that each value is a category that should be uniquely identified.

In either case, the user does not have a clear indication of the actual value based on the graphical display, even if relative values (gray scale) or at least the differentiation of values (color scheme) is apparent. In order to allow the user to resolve the actual values from the map, the **Tool Panel>>Cell Query** tool is available. Once this tool is activated, the user can click in any of the **Map Display** areas with the left mouse button. The value will be reported in the **Message Board**. It may be the case that the text posted to the **Message Board** can exceed the default width. To see the full text, the user can use the left mouse button to enlarge the **Message Board** by click-and-dragging any edge of that menu with the left mouse button.

For Windows users, the last window clicked on tends to move in front of all other windows or menus. This can be inconvenient when working with the **Cell Query** tool because the **Message Board** becomes obscured by the **Map Display** window. Moving the **Message Board** to a location that will not be obscured by the **Map Display** window can avoid this issue. The user can terminate the **Tool Panel>>Cell Query** tool operation by clicking with either the middle or right mouse button while the cursor is positioned in the **Map Display** window.

Hypsometry

A user can look for an elevation-related trend to the values for a GRID by using two **Map Display** areas to simultaneously display the DEM and the GRID of interest. Another approach is to use the **Pan/Zoom>>Composite** tool to tint the graphical display of the GRID by using a hillshaded version of the DEM. Still another method is the **Tool Panel>>Hypsometry** tool. This display removes the effect of X and Y coordinates from the reporting graphic, and provides a clearer display of the correlation between the area and elevation. It first subdivides the AOI into elevation bands, then determines the percentage of the AOI area within each band. The analysis is summarized as a percentage, accumulating upward from the lowest elevation band to the highest. An example of this is shown in figure 25*A*.

As an extension of this tool, a line (hypsometric curve) can be plotted for each zone in a zone map. The **Hypsometry** graph in figure 25*B* depicts an example output. In this example, Category 1 (the black line to the right) is found from low elevations up to approximately 3 750 meters of elevation.

A

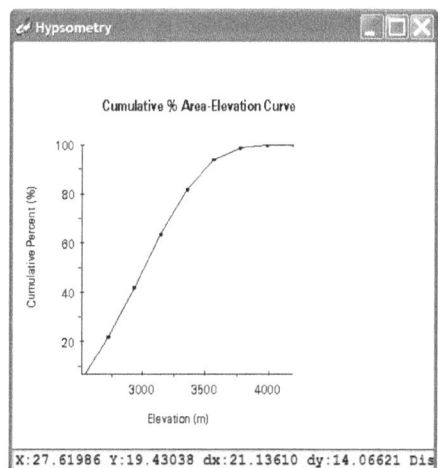

B

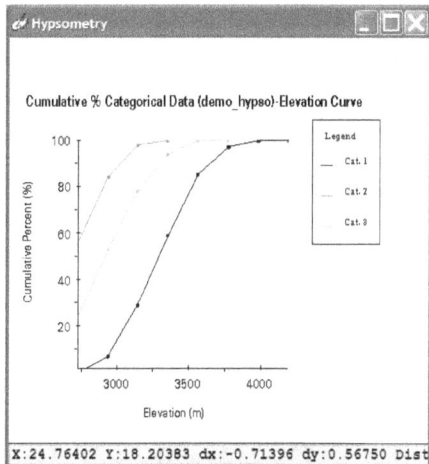

Figure 25. Example outputs from the **Tool Panel>>Hypsometry** tool for a *A*, single zone and *B*, three different zones.

Category 2 (the red line at the left) is exclusively found at lower elevations, 3250 meters or lower. The data used to generate this display are with the GIS Weasel software, located at .../weasel/dem and .../weasel/demo_hypso. This tool is only able to handle zone maps with eight or fewer zones.

Once the tool has completed executing, the **Hypsometry** graph will persist. The **Hypsometry** graph can be dismissed by clicking the **X** button at the top right corner of the graph.

Logic Query

Users can apply logical queries to any GRID that is available by using the **Tool Panel>>Logic Query** tool (fig. 23). A query is constructed by specifying a value from the **GRID** list, the **Operator** list, and the **Value** list, sequentially. Each choice is displayed in the **Query Logic** area at the top left of the menu. Figure 26 shows the **Query Logic** of *dem GE 3000*.

The **GRID** list shows the name of all the GRIDs in the Write Directory. If the desired GRID is located elsewhere, the user can click in the **GRID** list with the right mouse button to gain access to the **Grid Manager**. With the **Grid Manager**, the user can navigate to any directory on the hard drive and find their data. Once the GRID is specified, the name will be posted to the **Query Logic** area. Although only the GRID name and not the full pathname will be posted to the **Query Logic** area, the full pathname will be used when the query is actually processed.

The **Operator** list contains the standard logical operators:

GT – greater than
GE – greater than or equal to
LT – less than
LE – less than or equal to
EQ – equal to
NE – not equal to

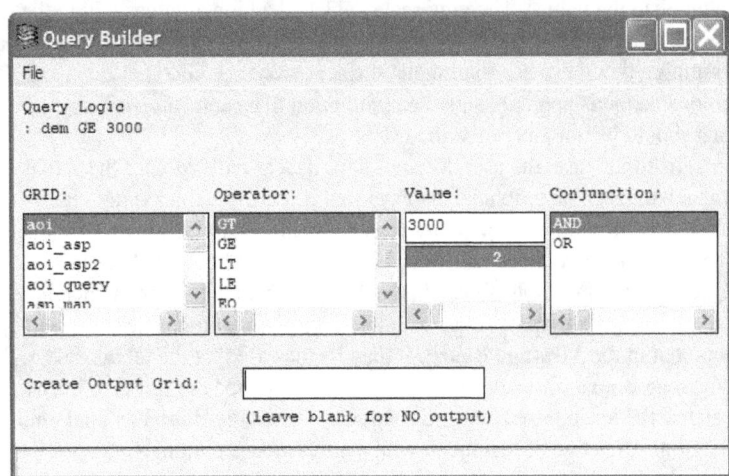

Figure 26.   The Query Builder Menu.

The **Value** input slot allows the user to specify the desired threshold by typing the value into the input slot. Alternatively, the value can be selected from the scrolling list below the input slot (shown by "2" in figure 26). This scrolling list is populated with the values of the VAT from the GRID highlighted in the **Grid** list, if it exists.

To construct a compound query (that is, multiple predicates), select the appropriate value (either "*OR*" or "*AND*") from the **Conjunction** list by clicking on the appropriate term and then return to the GRID list and begin the sequence again to create the query. If the query is not compound, then no selection from the **Conjunction** list should be made (that is, do not click on either choice).

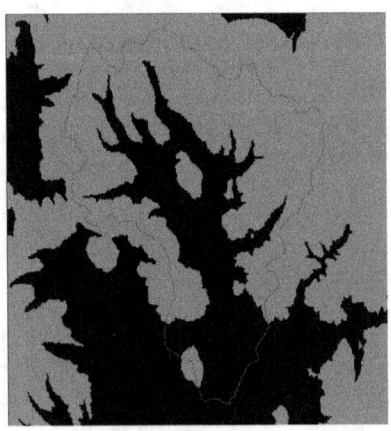

Figure 27.   Example of Tool Panel>>Logic Query output.

At the top left of the **Query Builder** tool is a pull-down menu labeled **File**. This menu can be opened by clicking on it with the left mouse button. The three choices available in the menu are *Submit Query Logic*, *Clear Query Logic*, and *Close Query Builder*. If *Submit Query Logic* is selected, then the query will be executed and the results will be drawn (in red) to the main **Map Display** area. An example of the display generated by the **Query Builder** tool is shown in figure 27. If there is a value specified in the **Create Output Grid** slot when the *Submit Query Logic* is selected, then the output will be written to a GRID with the name specified in the **Create Output Grid** slot. The GRID will be located in the Write Directory. Query logic cannot be edited. It can only be completely erased with the *Clear Query Logic* option and specified anew. *Close Query Builder* will terminate the **Query Builder** tool.

Delineation of Features

The GIS Weasel provides more than 20 different tools for creating new zone maps. These tools are separated into two functional groups: those that require little or no user input (Automated Methods) and those that require substantial user input (Interactive Methods). All tools require that the Tool Panel>>Current Zone Map be specified prior to being invoked, although some tools only use this information to spatially constrain the output. The user is encouraged to work through the "Sample Usage of the GIS Weasel" section, because it shows a large number of these functions and examples of the spatial data products that can be expected from these tools.

As mentioned in the following section and fully discussed in the "Advanced Topics" section, most of the products of these tools are automatically stored by the GIS Weasel. The user is given a copy of the final output in the Write Directory. The version managed by the GIS Weasel will be reused whenever a tool is rerun. In other words, each tool will check to see if the output already exists in the GIS Weasel Output Subdirectories (this is described in the "Advanced Topics" section). If so, the tool will simply allow the user to name a new copy of these data. This is important to understand if the user wants to run the tool "from scratch." To do this, the user is referred to the "Advanced Topics" section.

Overview on Output of Zone Maps

All the tools for delineating zone maps will produce a GRID that is stored in the Write Directory. The name of the GRID is specified by the user through the **Output Zone Map** menu (fig. 28). The user can type a new name into the slot to replace the default name that is presented. The name must be less than 12 characters. Alternatively, the user can click in the slot with the right mouse button to invoke the **Grid Manager**. From this menu, the name of a pre-existing GRID can be selected. The name cannot be *aoi*, as this is already used for the AOI GRID. The name cannot be the same as that of the Current Zone Map.

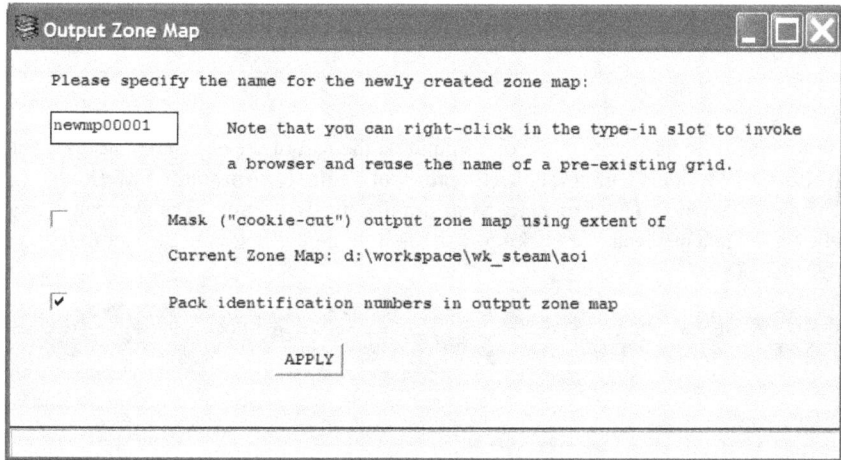

Figure 28.   The **Output Zone Map** menu.

The two checkboxes below the slot for specifying the name can usually be left in the default positions. The upper checkbox, whose label begins with **Mask...**, indicates whether the newly created zone map should be cropped or spatially constrained by using the Current Zone Map. It is important that the user understand this concept. A pair of examples will help illustrate the effect of this setting. First, if a drainage network is extracted from the flow accumulation surface that extends well beyond the boundary of the Current Zone Map, then the network will not necessarily be confined to it. By turning *On* the **Output Zone Map>>Mask...** checkbox, all links in the preliminary drainage network that lie beyond the boundary of the Current Zone Map will be discarded. A second example is when the user elects to derive the contributing areas associated with the links of a drainage network, the **Output Zone Map>>Mask...** checkbox should be turned *Off* because cropping the contributing areas down to only the places where the drainage network exists (that is, has nonnull values) will simply yield a map that has the exact same silhouette as the original drainage network.

Setting the lower checkbox, labeled **Pack identification numbers...**, to the *On* position ensures that the set of identification numbers associated with the zones will form a consecutive series. If the delineation tools yielded a preliminary zone map with three zones, whose identification numbers are [1, 3, 5], the **Output Zone Map>>Pack identification numbers...** checkbox in the on position would result in a map with the exact same zone shapes, but would be identified as [1, 2, 3]. The original sort order of the zones is preserved. Preliminary zone number 1 would still be numbered 1. Preliminary zone number 3 would now be numbered 2. Preliminary zone number 5 would now be numbered 3. It is a good idea to leave this option turned on for all cases, unless there is a specific reason to turn it off.

In many modeling softwares, identifiers are not explicitly associated with each feature. Rather, the identity of a specific feature is based on its position within one or more arrays that describe all instances of that type of feature. In other words, the position in the array, or *sort order*, defines the identity of the feature. The GIS Weasel does not track the identity of features on the basis of sort order, but frequently needs to produce information in this way for the benefit of the model that will consume the output. Therefore, to ensure that the sort order is identical to the identifier, the identification numbers should be packed.

In the case where the GIS Weasel is used to produce information about the connection (for example, flow routing) between two features, such as the identification of the downstream channel link, the information produced is the GIS Weasel identifier for the downstream link. For instance, if the identifiers for a set of channel links are [1, 3, 5], and link 1 flows into link 3 and link 3 flows into link 5, then the array of values describing the identity of the downstream link will be [3,5,0]. The first value will specify which link the first feature flows to, the second value specifies which link channel link 3 flows to, and the third value specifies which link channel link 5 flows to (this value is 0 because link 5 is the end of the network and has no down-slope neighbor). If a model uses these identifiers as synonymous with array position, then errors will occur. Outflow from link 1 will be erroneously given to link 5, because link 5 is in the third position in the array. The model will try to send outflow from link 3 to the link in the fifth position. Because there is no fifth position, the model will likely fail. This assumption is not true for all software that consumes GIS Weasel produced parameters. The user must understand how their chosen model works.

Although this topic will be discussed more completely in the "Advanced Topics" section, the storage of output from these methods will be introduced here. A delineation tool may yield a variety of intermediate GRID products. The user is not asked to name all intermediate products because they are automatically stored by the GIS Weasel. The GIS Weasel stores outputs in a hierarchy of subdirectories, which is generally referred to as the *Output Subdirectories* (described in the "Advanced Topics" section). These intermediate products are not destroyed because other methods may use these same intermediate data products. Further, the Output Subdirectories provides a metadata record on the choices made during a GIS Weasel processing session. If the user wishes to rerun a delineation tool (or parameterization method, described in "The Parameterization Phase" section) but wants to avoid previously created data products, then the "Advanced Topics" section should be read to understand what kinds of information are likely to be automatically reused by the GIS Weasel without informing the user.

Automated Methods

As mentioned, this grouping of tools requires little or no input on the part of the user. These tools are accessed by pressing the **Tool Panel>>Automated Methods** button. The **Delineation of Features: Automated Methods** menu (fig. 29) will be presented as a result. This menu also has a **Current Zone Map** slot like the **Tool Panel**. The user can change the Current Zone Map specification here by typing in a new value or clicking with the right mouse button in the slot to use a **Grid Manager**.

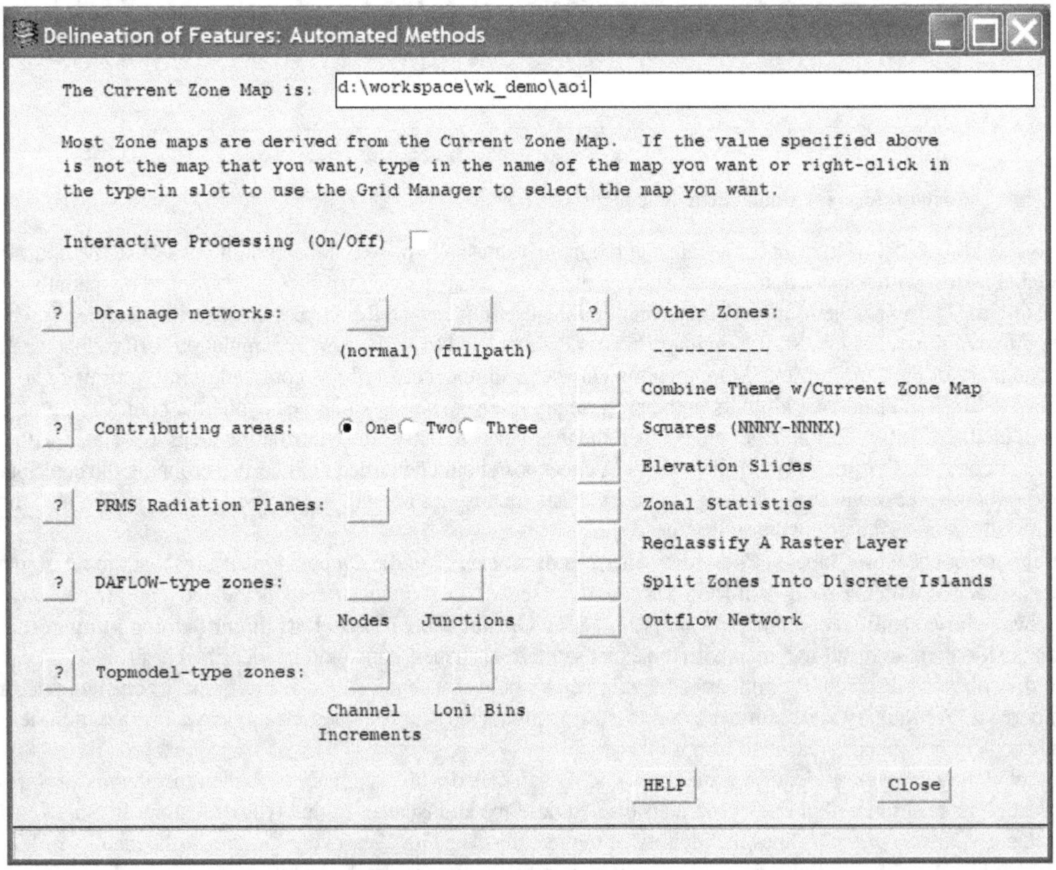

Figure 29.   The **Delineation of Features: Automated Methods** menu.

In addition, there is a checkbox for specifying whether or not to run these tools in the **Interactive Processing** mode. If so, then the user has the opportunity to change any default values that a tool might make available. The user is encouraged to turn *On* the **Interactive Processing** checkbox during the first usage of each delineation tool to gain an understanding of the methodology being used. To the left of most tool buttons is a button labeled with a question mark. These buttons will provide description of the associated tool. Pressing the **Help** button will display a general description of the different types of automated methods. More detailed information is accessible through the question mark buttons. The **Close** button will dismiss the **Delineation of Features: Automated Methods** menu without attempting to create a new zone map. The **Delineation of Features: Automated Methods** menu will terminate once the output is named.

Drainage Network: (normal) and (fullpath)

One of the most basic types of zone map for applications that rely on topographic information is the drainage network. The GIS Weasel provides two methods for delineating a drainage network, referred to as the *normal* and *fullpath* methods. Both of these methods find all the cells in the Flow Accumulation surface that exceed a user-defined threshold and designate the cells to be a part of the drainage network. The fullpath method will then perform further processing, which is described later.

The flow accumulation value for a cell is the number of cells upstream from that cell. Because each cell has an area, the flow accumulation value is an area value. The threshold value is the minimum area that is needed to support the initiation of a first-order link in the drainage network. Initiation refers to the upstream end of the link, as distinct from the downstream end. The **Drainage Extraction** menu (fig. 30) is invoked when either the **Drainage networks>>(normal)** or **Drainage Networks>>(fullpath)** button is pushed. This menu is used for collecting this threshold. The threshold will be applied by using the value shown in the **Cell Count** slot. This value, an integer, refers to the flow accumulation value. The size of the cell (length, not area) is shown in the text just above the **Threshold** area on the menu.

Although the **Drainage Extraction>>Cell Count** value is the figure that is ultimately used, there are a variety of other **Threshold** slots that are expressed in other units. These alternates are intended as a convenience for the user. After typing in a number into any of

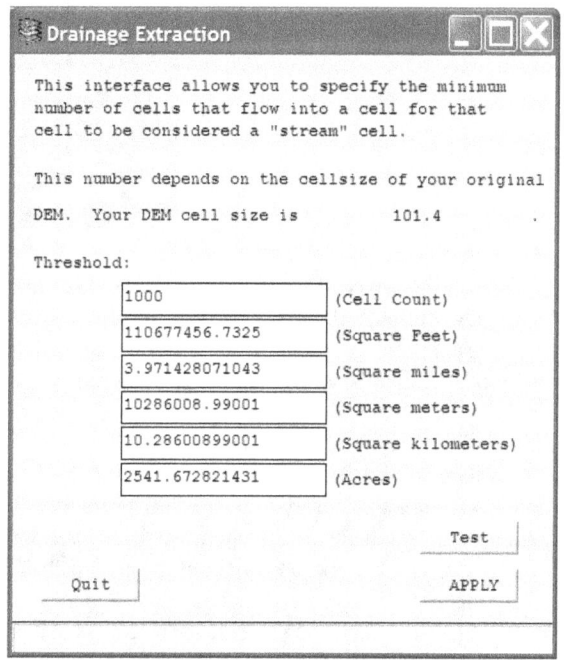

Figure 30. The **Drainage Extraction** menu.

the **Threshold** slots and pressing the <enter> key, all other slots are updated to equivalent values in the respective units. The **Cell Count** must be an integer, regardless of which slot was used. All alternate unit threshold values are rounded off to the nearest integer of **Cell Count**. For example, in figure 30, if the user entered 2542 in the **Acres** slot, it would be adjusted to the value shown in the figure.

There are three buttons on the **Drainage Extraction** menu. **Quit** dismisses the tool without attempting to create output. The **Apply** button will take the currently specified **Cell Count** value and delineate the drainage network. The **Test** button will create and display a temporary drainage network on the basis of the currently specified **Cell Count** value. The **Test** button is often an important interactive tool to help the user select the threshold value, because this value is difficult to anticipate (otherwise the GIS Weasel might have automated it.).

Figure 31 shows the effect of the threshold on the resultant drainage network derived from the sample DEM that is distributed with the GIS Weasel software. The display was generated by using the **Drainage Network>>Test** button, and shows a temporary version of the network. Once the **Drainage Network>>Apply** button is pressed, links are uniquely identified (which is not the case in figure 31). In figure 31*A*, the result of a relatively low threshold of 200 cells is shown (the DEM cell size is 100 meters). This means that the area upslope from the upstream end of every single first-order link is 200 cells. By substantially increasing the threshold to 5000 cells, a much-reduced drainage network density is created (fig. 31*B*). Users sometimes are tempted to use a very low threshold in order to push the starting point of first-order streams close to the drainage divide. A potentially negative side effect of this is the creation of many more links than is desired. Although there are other tools in the GIS Weasel that could be used to remove the spurious links, it can still be a relatively labor-intensive task. If the user wants a density like what is shown in figure 31*B*, but wants the starting points of the first-order links to begin at the drainage divide, then they should use the **Drainage Network>>Fullpath** method instead of the **Drainage Network>>Normal** method.

Figure 31. *A*, a drainage network with a cell count threshold of 200 *B*, a drainage network with a cell count threshold of 5000.

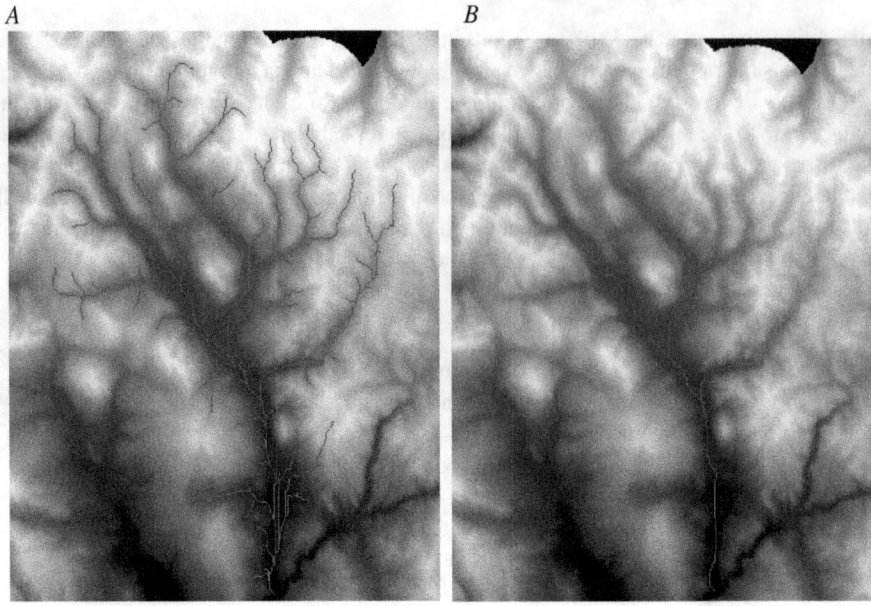

Figure 32 depicts the difference in the products of the **Drainage Network>>Normal** (fig. 31*A*) and the **Drainage Network>>Fullpath** (fig. 31*B*). A **Drainage Network>>Cell Count** value of 2000 was used in both cases. Both networks have the same density. The only difference is that the **Drainage Network>>Fullpath** method took what would have been the final product of the **Drainage Network>>Normal** method and extended the first-order links from the normal starting points uphill until the Flow Direction surface indicated that there were no more upslope neighbors. The algorithm for crawling uphill seeks to find the longest path from the link to the drainage divide.

Several words of caution are warranted here. Although the drainage divide is usually thought of as the "edge of the watershed," this is not a necessary characteristic for finding the starting point of a **Drainage Network>>Fullpath** link. The starting point may, in fact, be internal to the watershed. An example of this is shown in figure 32*B*, with a reddish-brown link found near the center of the network to the right of the main path. There is an elevation peak where this link begins, but this peak is not on the perimeter of the AOI (a watershed), shown in the red line. Another characteristic worth understanding is that sometimes the stream extension algorithm yields first-order links that have a "candy-cane" shape. An example of this is seen in figure 32*B* in the light blue/gray link found in the upper half of the AOI at its left edge. This occurs because the longest path will indeed come very close to the divide, but then seeks to climb an adjacent valley wall because of the increased length this entails.

Figure 32. An example of the output from *A*, the **Drainage Network>>Normal** tool and **B**, the **Drainage Network>>Fullpath** tool.

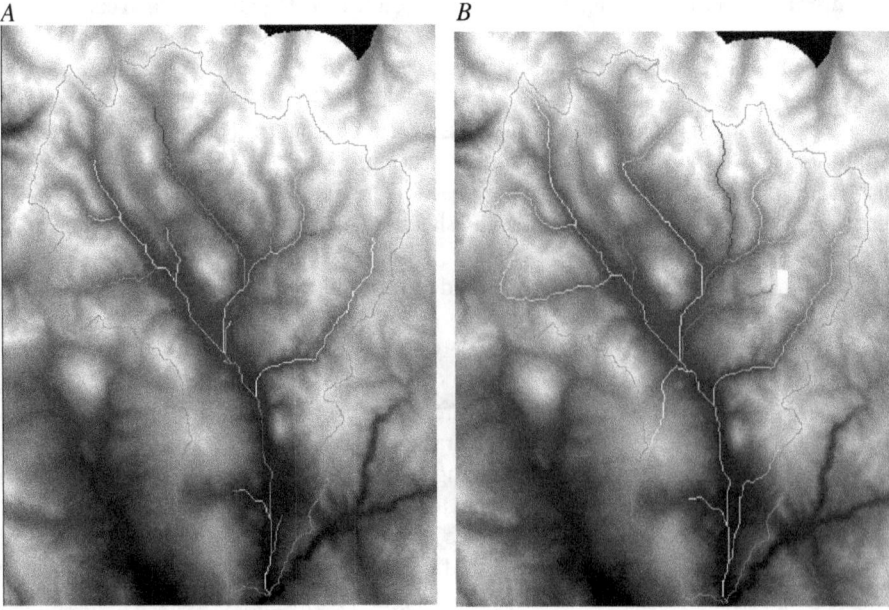

Interactive Processing

If the user turned *On* the **Delineation of Features: Automated Methods>>Interactive Processing** checkbox prior to pressing one of the **Delineation of Features: Automated Methods>>Drainage Network** buttons then the user will be asked whether to create a **Threshold Table**, an example of which is shown in figure 33. This table gives estimates of the number of links that can be expected in the drainage network (derived from the Input Elevation Grid) at a variety of thresholds. A link is defined here as the line between the headwater point of a first-order stream and the confluence with another link, the line between confluences (for higher-order links, or from the downstream-most confluence and the outlet of the AOI (for the downstream-most link). These numbers are approximate; the result may have a different number of links because of the effect of the **Output Zone Map>>Mask...** checkbox (described earlier).

Problems and Limitations

These drainage extraction methods are relatively simple. There are other, more complex methodologies described in the literature. Regardless of which method is used, the output is only going to be as good as the input. The quality of the derived drainage network is a clear indicator of the quality of the Input Elevation Grid. Unfortunately, the GIS Weasel does not offer tools for modifying the elevation data. There are tools available for this purpose that the user is free to employ prior to beginning the GIS Weasel processing session.

In general, the delineation of drainage networks is difficult in flat areas. This becomes obvious when a lake or reservoir is present because the drainage lines that cross it tend to be perfectly straight. This is caused by the fact that in a perfectly flat area, there is no difference in elevation between neighboring cells and the flow direction algorithm has no further information on which to resolve meanders. The algorithm will simply connect each inflow location to the flat area to one of the outflow locations by using the most direct route possible. In addition to straight paths, the selection of which outflow location by the flow direction algorithm is arbitrary and may not be a correct reflection of what is on the landscape.

Very rough topographies can lead to drainage networks that are perceived as too sinuous, but more likely, because of filling the depressions in the DEM (described in "The Setup Phase" section), will lead to drainage networks with a low degree of sinuosity. By filling a broad, hummocky area, the elevation surface may be raised to a consistent peak height of the hummocks, resulting in a flat area. This yields a similar condition to what was described in the preceding paragraph.

Cell Threshold	# of links in network
0	0
10	4775
25	1861
50	840
100	356
200	202
300	119
400	82
500	66
600	53
700	44
800	40
900	35
1000	32
1100	28
1200	28
1300	26
1400	26
1500	26
1600	26
1700	26
1800	26
1900	26
2000	26
2200	22
2400	18
2600	16
2800	16
3000	14
4000	10
5000	6
10000	4
20000	2

Figure 33.   The **Threshold Table**.

Another common morphology of the delineated drainage network that tends to catch the eye of the user is where several drainage lines come into close proximity to and run parallel with each other, sometimes in immediately adjacent cells, prior to converging. The human interpreter usually feels that the convergence should occur upstream from where the delineation specifies, at the location where the drainage lines first begin to run parallel to each other. This "parallelism" is again the artifact of the flow direction algorithm having trouble resolving what to do in flat areas. These parallel lines occur almost exclusively in valley bottoms.

There is a minimum link length within the drainage network. A link can be as short as a single cell. Such a link sometimes occurs when the DEM indicates that two confluence points are separated by a very short distance; a link is delineated for this inter-confluence distance. In reality, the three streams associated with these confluence points may converge at the same confluence point. Although this begs the question of what is a "point" (at what scale does the confluence point become an area?), replication of the confluence of three links at the same cell is, in practice, rare. The user is encouraged to use the **Pan/Zoom** menu to more closely examine areas where this may (or may not) be occurring.

Before the user resorts to expending substantial effort modifying their Input Elevation Grid to avoid these problems, they should reflect on how these kinds of artifacts will effect their application. The term *drainage network* has been purposely chosen here to avoid encouraging, as much as possible, the user to attempt to recreate the *stream* or *river* network. Although such word-smithing has a minimal effect, it hopefully reminds the user that the drainage network is being constructed strictly for the purposes of the modeling effort. Recreating a cartographically or hydrologically accurate representation of streams as found in the field is by no means required for this. It simply may not be possible, let alone profitable, to find the precise threshold that ensures the accurate initiation of headwater stream segments across a watershed. In the end, the user needs to understand whether the environmental simulation model being used will be sensitive to these issues. For example, the parameters by which many models gain understanding of the reality being simulated may not even have the capacity to indicate where the streams begin or how long they are. The GIS Weasel obviously will not have this model-specific expertise and, therefore, relies on the judgment of the user.

There are several other tools (**Tool Panel>>Delineation of Features>>Drain fromPoint**, **Tool Panel>>Digitize**, **Tool Panel>>Delineation of Features>>Id # Manipulation**) that are useful for modifying a "first-cut" delineation produced by the methods currently being discussed. These can be used to "clean up" the drainage network map. For instance, some users will use the **Drainage Network>>Fullpath** tool, and then use some of the other tools just mentioned to eliminate the "candy cane" effects that, when they are present, by adding new links, removing spurious ones, and reassigning identification numbers to the zones to group the new link with a pre-existing one.

In some cases, the effect of the quality of the drainage network on a user's specific application is a function of limitations in the raster encoding scheme (that is, GRID) used to represent the drainage network. In reality, streams have a width, as well as an area. Further, streams have depth that varies across the width of the stream (sometimes referred to as the cross-section). A raster drainage network from a DEM should generally not be used to make inferences about stream width or cross-section shape or area. Bed roughness is another much sought after characteristic that the DEM is not able to reveal. Put another way, the DEM and the derivatives are usually not appropriate for resolving the types of characteristics relevant to hydraulic engineering, stream habitat assessment, or other studies requiring in-channel characterizations. A GRID has a cell size that, despite ever improving computing abilities and DEM quality, is generally much coarser spatially than what the in-channel characterizations require. For example, stream width can only be described in increments of DEM cells and must be at least one cell. Therefore, if a DEM has a cell size of 30 meters, the stream can be no narrower than one cell (30 meters).

The user is warned to avoid bringing an externally created drainage network GRID into a GIS Weasel processing session that is not consistent with the DEM. This may be tempting to do if the user's DEM is of poor quality and the DEM-derived drainage networks consistently fail to represent what is correct or needed. The reason that this inconsistency can cause problems is that if any topographic derivatives are made from the non-DEM based drainage network, these derivatives will not be based on just the drainage network, but also the DEM or the derivatives. Examples of derived information that may be negatively effected by using a non-DEM based drainage network are the delineation of contributing areas (some areas may not flow to any link) and the determination of the downstream link in the drainage network (the flow direction at the outlet cell of a link may not point to the adjacent cell representing the start of the downstream link).

Contributing Areas: One, Two, Three

If the user would like to find the area or areas upslope from one or more zones in their Current Zone Map, they can use the **Delineation of Features: Automated Methods>>Contributing Areas** tools. The three tools use the Flow Direction surface, so again the quality of the output is dependent on the quality of the Input Elevation Grid and the ability to resolve flow direction from it. The input to these tools is usually a drainage network, but this is not a requirement. If the user is interested in determining the area above zones representing land use, vegetation patches, or man-made structures (such as roads or bridges), all that is needed is a GRID depicting these features. The zones within this Current Zone Map can be as small as a single cell or composed of many cells. If there are multiple cells, there is no limitation on the shape that the zones form.

The example outputs (fig. 34) are based on the sample DEM that is distributed with the GIS Weasel software, and a drainage network extracted from it using a threshold of 2000 cells. The three tool buttons are labeled:

- **Delineation of Features: Automated Methods>>Contributing Areas>>One**
- **Delineation of Features: Automated Methods>>Contributing Areas>>Two**
- **Delineation of Features: Automated Methods>>Contributing Areas>>Three**

The products of these three tools are found in figures 34*A*, 34*B*, and 34*C*, respectively. The lines of the drainage network have been superimposed onto the output product and are not actually integrated into the output product. For ease of reference, the outputs from these tools are referred to as *one-plane*, *two-plane*, and *three-plane* zone maps, respectively, referring to the number of contributing areas associated with each zone in the Current Zone Map. Each is more fully described in a correspondingly named section.

A

B

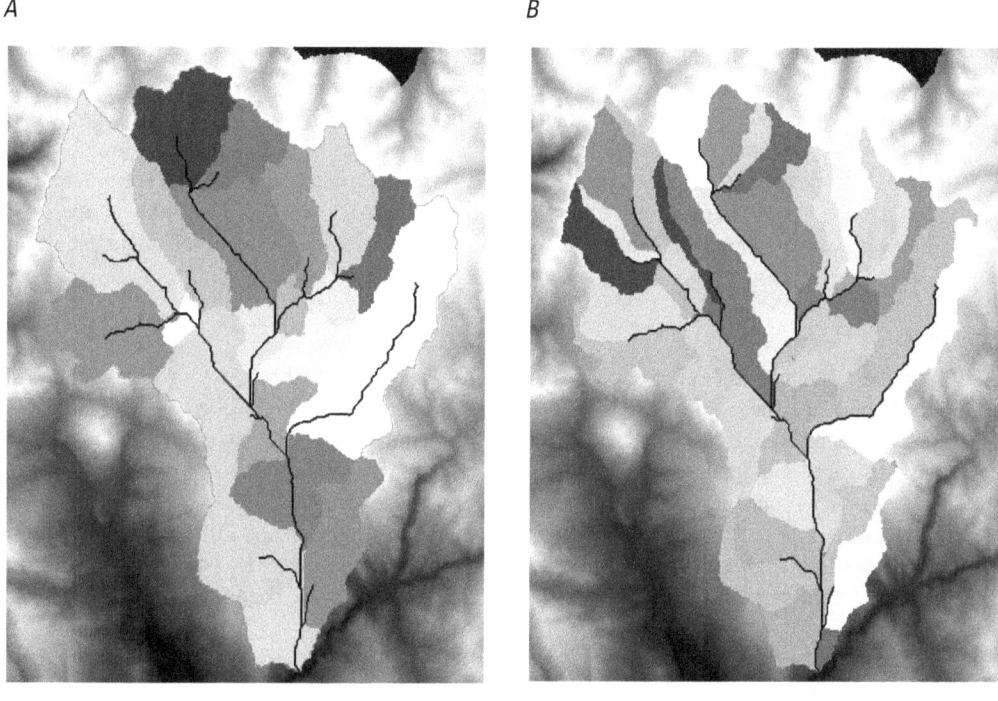

C

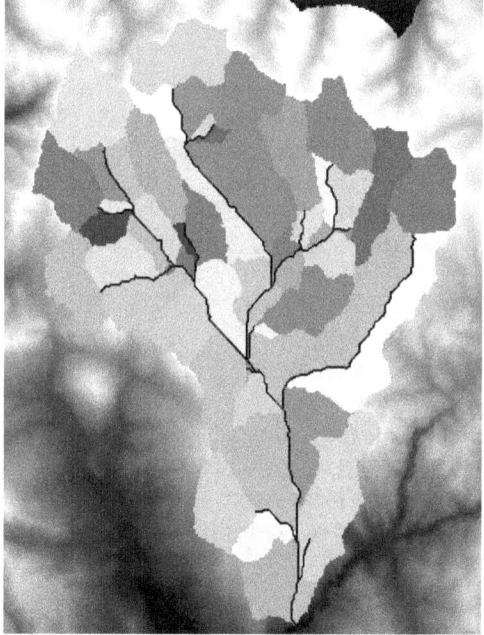

Figure 34.   Examples of the output from the *A*, **Contributing Areas>>One** *B*, **Contributing Areas>>Two** *C*, **Contributing Areas>>Three** tools.

One Plane

Delineation of Features: Automated Methods>>Contributing Areas>>One will produce a single zone upslope from each zone in the Current Zone Map. This zone can be thought of as a watershed or basin. The output from this tool will be stored in ...*output/zones/<Current Zone Map>/one-plane/grid*, where *<Current Zone Map>* is a reference to whatever value is specified in the **Delineation of Features: Automated Methods>>Current Zone Map** slot when the tool is run. The user will be prompted to name a copy of this output that will be stored in the Write Directory.

All the cells that flow into a zone in the Current Zone Map will be assigned the value within the one-plane zone map that matches the identification value of the source zone in the Current Zone Map. The output zone will include cells that are located within the source zone of the Current Zone Map. For example, if a drainage network is the Current Zone Map, the output one-plane map will not have a series of null-valued cells where the cells that form the links in the drainage network lie. Further, those cells that are located in the position of the zones (that is, the links) in the Current Zone Map are not differentiated from the enclosing contributing areas. The graphic in figure 34*A* shows the drainage network map, but this is only a superimposition of a vector version of the drainage network onto the one-plane GRID to visually emphasize to the reader where the original drainage network links were located.

Two Planes

Delineation of Features: Automated Methods>>Contributing Areas>>Two will produce a pair of zones upslope from each zone in the Current Zone Map. These pairs of zones form a left and right bank to the zone in the Current Zone Map. This can be important in applications where incidence of solar radiation is important because the two areas tend to have contrasting aspects. This separation also can be important for various kinds of algorithms for routing water.

The tool will take the one-plane derivative of the Current Zone Map (stored at ...*output/zones/<Current Zone Map>/one-plane/grid*) and bisect each of those zones by using a drainage network derived from the Current Zone Map (stored at ...*output/zones/<Current Zone Map>/fullpath2/grid*). The *fullpath2* product ensures that first-order links extend to the topographic divide so that the one-plane zones for these links can be bisected. If the Current Zone Map already depicts a fullpath-type drainage network, no ill effect will result.

This tool is more specifically engineered to exploit a Current Zone Map that represents a drainage network. If the Current Zone Map is not a drainage network, then the fullpath2 product will attempt to determine the main line of drainage through the zone. The ultimate output from this tool will be stored in ...*output/zones/<Current Zone Map>/two-plane/grid*, where *<Current Zone Map>* is a reference to whatever value is specified in the **Delineation of Features: Automated Methods>>Current Zone Map** slot when the button is pushed. The user will be prompted to name a copy of this output that will be stored in the Write Directory.

As noted at the end of the explanation of the one-plane tool, the cells constituting the zones in the Current Zone Map are always assigned to a zone in the output. In the case of the one-plane map delineation, assignment of the cells in the Current Zone Map to a zone in the output zone map is straightforward. In the case of the two-plane delineation, it is less straightforward because the cells of the Current Zone Map are actually a boundary between the two planes. Rather than exclude the boundaries from both zones in the resultant map, these cells are (on a per-cell basis) arbitrarily assigned to either the left or right bank zone.

There could be a single (that is, only a left or a right bank) zone associated with a zone in the Current Zone Map. This can occur for zones that are very short links in a drainage network, because all the cells near to it are associated with a different link. This also can happen even if the link is quite long, but runs next to the drainage divide. An example of this is a stream at the bottom of a cliff wall, where rain onto the cliff top flows to a different link in the drainage network. Yet another reason that a only a single zone might be delineated is if the fullpath2 product was unable to extend a first-order drainage link to a topographic divide, or extended it to a topographic peak that is not on the perimeter formed by the topographic divide. This would effectively cause the failure of the bisection of the one-plane zone (the fullpath2 link would not go from edge-to-edge).

Sometimes there are actually more than two zones in the output for each zone in the Current Zone Map. This usually happens when groupings of just a few cells end up being spatially isolated from a larger group of cells with which the small group should be lumped. This is a fairly common artifact of delineating boundaries within a GRID. Fortunately, the GIS Weasel provides tools for cleaning up these artifacts if the user desires. While spurious zones may be detected by visually inspecting the output in the **Map Display** areas, it is often difficult to spot very small zones. A useful tactic for finding these spuriously small zones is to use the **Pan/Zoom>>Raster Control>>List Table** tool to look at the VAT for the input map (that is, the drainage network) to determine the number of links and then to look at the VAT for the newly derived two-plane type output.

Another case is when too many zones in the output (again, the user should understand whether this is "too many" for their specific application) or there are zones that are smaller than is considered appropriate. In these cases, **Tool Panel>>Dissolve** can be used to automatically group small zones into larger neighbors, and **Tool Panel>>Id # Manipulation** can be used to interactively reassign the identification number of a zone (usually to match that of legitimate zone). The operation of these two tools is discussed later. With regard to the **Tool Panel>>Dissolve** tool, the user will be asked for the minimum allowable area. This number is often decided upon while using the **Pan/Zoom>>Raster Control** tool to examine the VAT of the initial two-plane map, which will show the *COUNT* of the number of cells within the zone.

Three Planes

Delineation of Features: Automated Methods>>Contributing Areas>>Three produces a map that is almost identical to that produced by the **Delineation of Features: Automated Methods>>Contributing Areas>>Two** tool, except that the area above the upstream end of the first-order links associated with the Current Zone Map is given a one-plane type of zone (that is, a watershed above the start of the first-order links), instead of being split into right- and left-bank areas. The one-plane derivative of the Current Zone Map will be an intermediate product (stored at *...output/zones/<Current Zone Map>/one-plane/grid)*. There is no fullpath product. The ultimate output from this tool will be stored in *...output/zones/<Current Zone Map>/three-plane/ grid*, where *<Current Zone Map>* is a reference to whatever value is specified in the **Delineation of Features: Automated Methods>>Current Zone Map** slot when the button is pushed. The user will be prompted to name a copy of this output that will be stored in the Write Directory.

PRMS Radiation Planes

This is a specialized tool developed to support the USGS Precipitation-Runoff Modeling System (PRMS, *http://wwwbrr.cr.usgs.gov/mms*), a surface-water model. The GIS Weasel also provides support for generating many PRMS-specific parameters. These are documented in the "Parameterization Methods" appendix.

The expected type of Current Zone Map is a two-plane map (see previous section), although this is not required. The PRMS radiation plane map is created by first reclassifying the slope (*...output/surfaces/<Input Elevation Grid>/filled/slope/reclass/ grid*) and aspect (*...output/surfaces/<Input Elevation Grid>/filled/aspect/reclass/grid*) surfaces and then combining them into a radiation surface (*...output/surfaces/<Input Elevation Grid>/filled/radpl/grid*). The *filled* token in the pathnames will not be present if the user declined to fill the pits in the Input Elevation Grid. The most commonly occurring value of the radiation plane surface within each zone of the Current Zone Map is then assigned to all the cells in that zone in the output. The output from this tool will be stored in *...output/zones/<Current Zone Map>/radpl/grid*, where *<Current Zone Map>* is a reference to whatever value is specified in the **Delineation of Features: Automated Methods>>Current Zone Map** slot when the button is pushed. The user also will be prompted to name a copy of this output that will be stored in the Write Directory.

The VAT for this output will contain extra information. Normally, the only items in VATs are *VALUE* and *COUNT*. In anticipation of using this output in support of PRMS, extra information has been added, the most important of which are *SLOPE* and *ASPECT*. The other items are used for processing but are not produced as output.

Figure 35 shows an example of the radiation plane map, based on the zone map shown in figure 34*B*. Although it has the same shapes as the zones in the Current Zone Map, the colors, signifying the identification numbers, are different. These colors correspond to the dominant (most commonly occurring) combination of reclassified slope and aspect within the zones in the input. Several colors, such as purple, recur throughout the map. These areas, which were uniquely identified in figure 34*B*, share the same identification number in the radiation plane map. The number associated with each cell indicates the type of radiation plane, as opposed to the identity of a unique instance. This overlap frequently occurs, but is not guaranteed to occur.

There is a single-celled zone at the center of the new map. This should not be removed. It is used as a (perfectly flat) reference radiation plane by PRMS.

If the **Delineation of Features: Automated Methods>>Interactive Processing** checkbox is in the *On* position, an informational menu describing the radiation plane derivation process in detail will be presented. This menu also has an **Interactive Processing** checkbox. If this is on, then the user will be able to modify the default settings for the reclassification of the slope and aspect maps, if these have not already been calculated within the Write Directory.

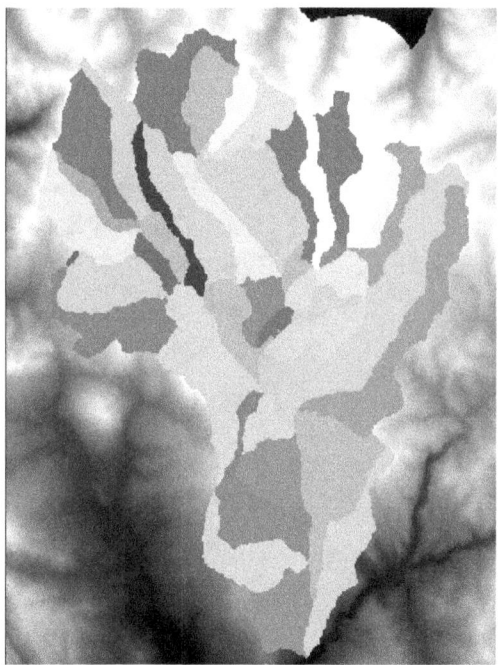

Figure 35. An example of output for the **PRMS-type Zones>>Radiation Planes** tool.

DAFlow-type Zones

DAFlow is a USGS model for simulating streamflow. The following two tools are used for delineating features related to the drainage network for the DAFlow model. Further documentation on the DAFlow model can be found in the DAFlow manual (Jobson, 1989). The GIS Weasel also provides support for generating DAFlow-specific parameters. These are documented in the "Parameterization Methods" appendix.

Nodes

Nodes are segmentations of a drainage network used to understand how water moves between the streambed to the ground-water system. The ground-water system is typically conceptualized as a finite-difference mesh like the one shown in figure 36. The size of squares in the mesh vary according to the application. The extent may extend to well beyond the AOI used for surface-water modeling, depending on modeling objectives. Figure 37*B* shows the DAFlow nodes generated based on the drainage network shown in figure 37*A* (which used a **Drainage Network>>(normal)>>Threshold>>Cell Count** of 2000) and the ground-water mesh of figure 36 (shown in red using the vector representation). In addition to differentiating new zones wherever a confluence occurs, a new zone identification number is assigned when a drainage link crosses a boundary within the ground-water mesh. Further, the cell that is the starting point for each link is differentiated with a new identifier, as seen by the green cell (forming a junction between the two tributaries starting at the top of the image) in figure 37*B*.

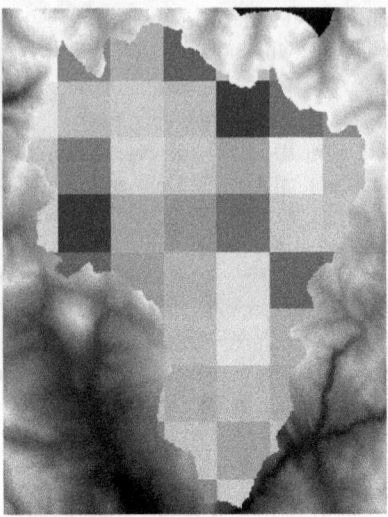

Figure 36.   An example of a finite-difference mesh used for ground-water modeling.

The output from this tool will be stored in ...*output/zones/<Current Zone Map>/ndanode/grid*, where *<Current Zone Map>* is a reference to whatever value is specified in the **Delineation of Features: Automated Methods>>Current Zone Map** slot when the button is pushed. The user will be prompted to name a copy of this output that will be stored in the Write Directory.

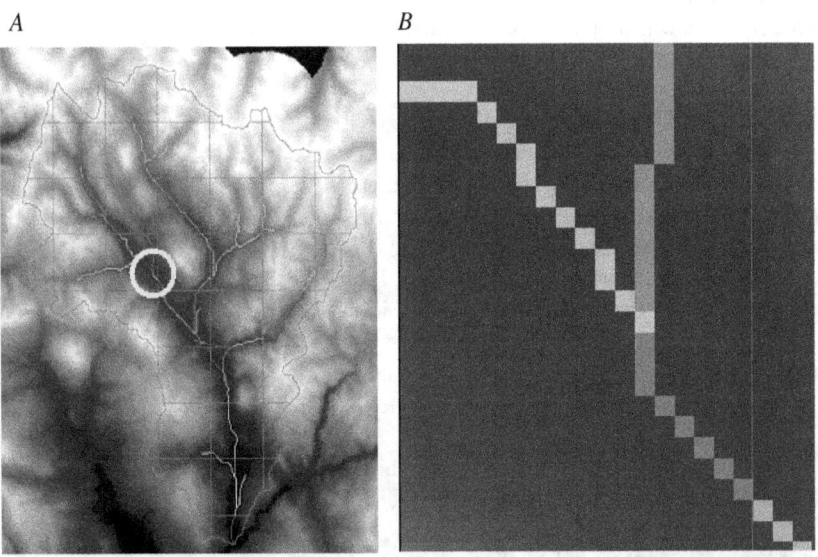

Figure 37.   A, an example of output for the **DAFLOW-type Zones>>Nodes** tool. B, a close-up view of a portion of the output, near the yellow circle in figure 37A.

Junctions

DAFlow junctions are simpler than DAFlow nodes. Junctions are simply the starting cell of each link in a drainage network. Strictly speaking, junctions are supposed to represent a point between the links in a drainage network, but this cannot be represented within a raster dataset. Therefore, as a compromise, the starting cell of each drainage link is used. Figure 38 shows sample output of this tool, again based on the drainage network shown in figure 32A. Figure 38*B* shows a close up of the junctions because junctions are so difficult to see at the full scale of the AOI (shown in red lines) in figure 38*A*. In both images, a vector version of the drainage network is superimposed on the display. The drainage network is not part of the output.

The output from this tool will be stored in ...*output/zones/<Current Zone Map>/ndajunction/grid*, where *<Current Zone Map>* is a reference to whatever value is specified in the **Delineation of Features: Automated Methods>>Current Zone Map** slot when the button is pushed. The user will be prompted to name a copy of this output that will be stored in the Write Directory.

Topmodel-type Zones

Topmodel is a surface-water model that is in wide use. The GIS Weasel provides two delineation tools to support the use of this model. The GIS Weasel also provides support for generating Topmodel-specific parameters. These are documented in the "Parameterization Methods" appendix. For more information on the operation of Topmodel, the user is referred to the documentation (Beven and others, 1995). It should be noted that there also is a plug-in to the GIS Weasel that provides support for a modified form of Topmodel, called WEBMOD (Webb and others, 2004). Plug-ins are introduced in the "Advanced Topics" section, although documenting each plug-in is beyond the scope of this manual. Each plug-in is expected to provide documentation.

Standard Topmodel is generally based on one-plane types of zones as the basic modeling response unit, although this has begun to change as the model has been extended (for example, WEBMOD). The examples of output shown here will be based on the one-plane zones shown in 34*A* as the modeling response unit. The modeling response unit often has been referred to as a *subcatchment*.

Channel Increments

Channel increments are intended to specify points along the main stem of drainage within each subcatchment. Fluxes of water from the surrounding subcatchment are added to the drainage network for routing at these locations. The input (Current Zone Map) to this tool is usually the subcatchment (that is, one-plane) map. The **Channel Increments** tool will dynamically determine the path of the main stem of drainage within each zone. A drainage network could be used as the Current Zone Map and control the determination of the "main stem." This tactic can be useful if the DEM is of poor quality. Each link in the drainage network effectively represents the corresponding subcatchment within the execution of the **Channel Increments** tool.

Once the main stem of drainage is found for each subcatchment, and if the **Delineation of Features: Automated Methods>>Interactive Processing** checkbox is in the *On* position, the menu shown in figure 39 is presented. This menu allows the user to specify how many increments, or sampling points along the main stem of drainage, there should be within each subcatchment. Each subcatchment can have a different number of increments.

The menu lists three pieces of information for each subcatchment: the subcatchment identification number (first column), the number of cells constituting the main stem (main link) of drainage within the subcatchment, and the default number (#) of channel routing increments for that subcatchment. The second column of information is important because it is preferred that there be at least 10 cells for every channel increment. In the case of a very short (that is, a small numbers of cells) main stem,

A

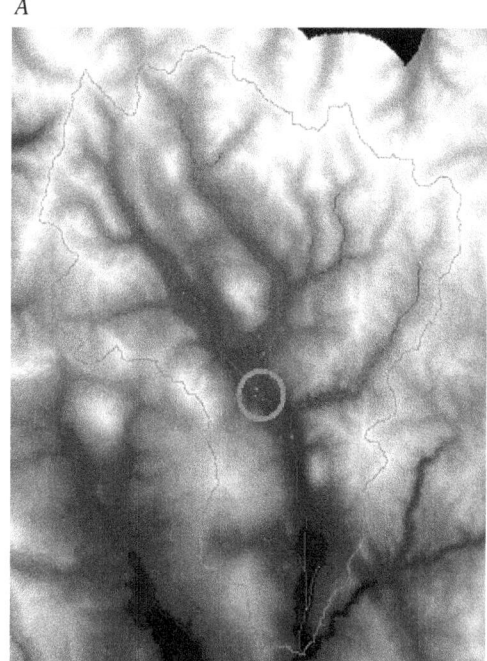

B

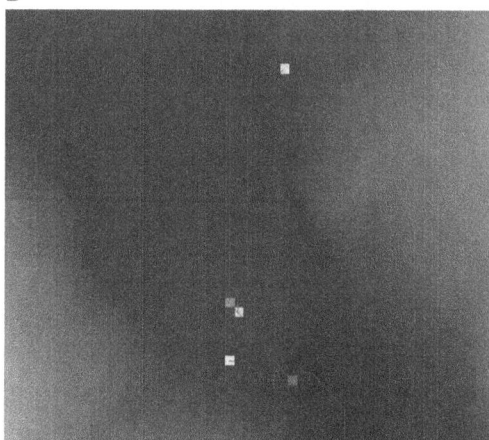

Figure 38.   A, An example of output for the **DAFLOW-type Zones>>Junctions** tool shown with the drainage network it was derived from. B, close-up of the area near the red circle in red on Figure 38A.

this can become a problem. To change the number of channel routing increments for a zone, click on the relevant row in the table and enter a new value in the resultant pop-up menu.

The default values for the number of channel increments is five. This is automatically adjusted when a short main stem of drainage is detected. Figure 39 shows that some default values for the number of channel increments is less than five. Generally, at least two channel increments are needed. This allows Topmodel to differentiate the volume of water in the stream at the upstream end of the subcatchment and the volume of water in the downstream end of the subcatchment, and thereby, exploit the conceptualizations about the variable source area for runoff generation within a subcatchment.

If the user discovers links that are too short during this process, then the process should be completed, but the results discarded and then return to the delineation of the subcatchments (or drainage network) to deal with this problem. Once this has been addressed, this tool should be run by using this amended subcatchment map.

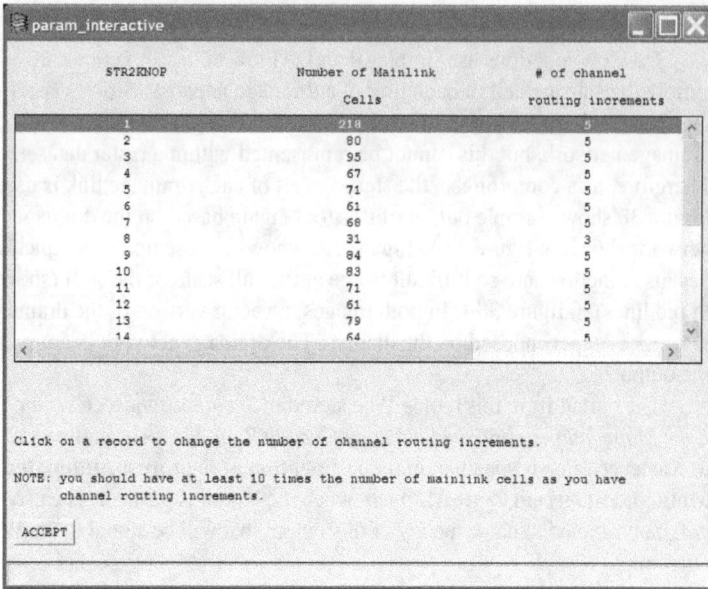

Figure 39. The menu for specifying the number of channel increments per subcatchment.

Figure 40 shows an example of the output from the **TOPMODEL-type Zones>>Channel Increments tool.** The red lines are superimposed vectors representing the subcatchment map and are not part of the product. Also note that the increments are shown here as points for better visualization. The output from this tool will be stored in …*output/zones/<Current Zone Map>/ntop-chan/grid*, where *<Current Zone Map>* is a reference to whatever value is specified in the **Delineation of Features: Automated Methods>>Current Zone Map** slot when the button is pushed. The user also will be prompted to name a copy of this output that will be stored in the Write Directory.

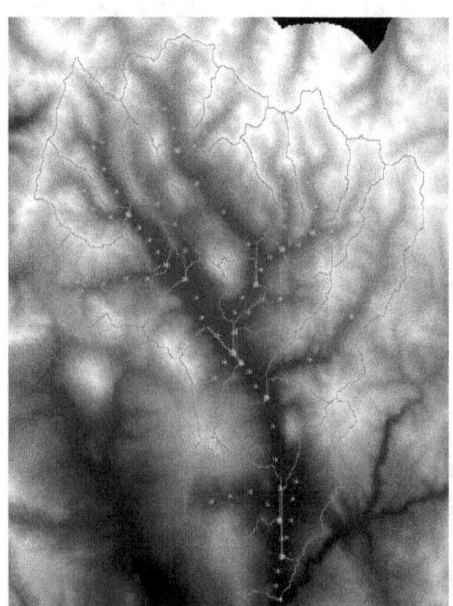

Figure 40. An example of output for the **TOPMODEL-type Zones>>Channel Increments** tool shown with the subcatchment it was derived from.

The Topographic Wetness Index and Loni Bins

Topmodel uses the Topographic Wetness Index (TWI) to characterize the likelihood of an area to become saturated. This also has been described as a relative indicator (to other parts of the subcatchment) of how close the groundwater table is to the surface. The user is referred to the Topmodel documentation for a discussion of the formulation of the TWI. Roughly, the TWI is the natural logarithm of the quantity derived by dividing the flow accumulation by the slope. The TWI is referred to here as the *loni*, after the symbol for natural log ("ln").

A subcatchment is constituted of a number of cells, each of which can have a loni value. Topmodel, rather than ingesting the entire set of loni values for each subcatchment, ingests a frequency distribution of the loni values within each subcatchment. To do this, the loni surface needs to be reclassified into bins. The number of cells (that is, frequency of occurrence) of each bin within a subcatchment is then described to the model. The generally preferred number of bins within a subcatchment is 30, although this is only a rule of thumb. The reclassification of the loni surface into the number of bins set for a subcatchment is based on the range and distribution of values in the loni surface within each subcatchment. This means that each subcatchment essentially has a unique reclassification of the loni surface values.

If the **Delineation of Features: Automated Methods>>Interactive Processing** checkbox is in the *On* position when the **Delineation of Features: Automated Methods>>Topmodel-type zones>>Loni Bins** button is pressed, then the user will be asked to respond to a variety of questions from the GIS Weasel. If this checkbox is in the *Off* position, the user will not be prompted to make any choices. The standard Flow Direction and Flow Accumulation surfaces will be used to derive the loni surface (stored at *.../output/surfaces/<Input Elevation Grid>/filled/loni/grid*) and the number of loni bins (stored at *...output/zones/<Current Zone Map>/loni-nbins/grid*) per subcatchment will default to 30.

In the interactive mode, a menu giving many details about the derivation of loni bins is first presented by the GIS Weasel. This menu also has an **Interactive Processing** checkbox. If this is in the *On* position when the **Okay** button is pressed, then the user is asked, possibly multiple times, whether the Flow Direction surface should be reused. The user should answer *Yes* to this question.

The user is then asked how Flow Accumulation should be derived for the purpose of deriving the loni surface (menu not shown). The top option refers to the standard Flow Accumulation that the GIS Weasel normally uses. The remaining two methods have specialized logic to adjust the effect of the contribution of flow from upstream subcatchments on the loni values of downstream subcatchments. Which is the correct choice is a matter of debate among modelers, and this is beyond the scope of this manual.

SC-ID	Count	Min-Loni	Max-Loni	Mean-Loni
1	9033	3.6174	21.3403	7.4798
2	4781	3.5209	20.6186	7.7874
3	3385	3.5818	20.2973	8.1131
4	2132	3.5209	19.8967	7.2386
5	4218	3.9000	20.5811	8.3289
6	3733	3.9392	20.4590	8.6645
7	2222	3.7406	19.9237	8.0251
8	2103	3.7406	19.8227	7.9912
9	4183	3.3354	20.5644	7.3121
10	3643	3.5209	20.3801	7.2621
11	4458	3.3354	20.6115	7.3885
12	2314	3.1604	19.9720	7.0407
13	2494	3.1789	20.0556	7.2158
14	2899	3.3354	20.2061	7.1990

Figure 41. Informational menu showing the per-subcatchment statistics of the loni surface.

STR2KNOP-ID	Count	# of Loni Categories
1	9033	30
2	4781	30
3	3385	30
4	2132	30
5	4218	30
6	3733	30
7	2222	30
8	2103	30
9	4183	30
10	3643	30
11	4458	30
12	2314	30
13	2494	30
14	2899	30

Click on record to be modified...

ACCEPT

Figure 42. The menu for specifying the number of loni bins per subcatchment.

Figure 41 shows a noninteractive (informational) menu that is presented to the user in conjunction with the interactive menu shown in figure 42. Figure 41 shows information about the distribution of the loni surface values within each subcatchment. This information is provided to the user as a way to look for problems in the derivation of the loni surface and to make decisions about how many loni bins are needed for each subcatchment.

Figure 42 shows a table displaying the subcatchment identification number (first column), the cell count (that is, indication of the area) in the subcatchment, and the number of loni bins (categories) to derive for each subcatchment. To change the number of loni bins for a subcatchment, the user can click on the relevant row in the table. A small menu with an input slot is presented. The user can type in a new number of loni bins. The table is then updated. When the user presses the **ACCEPT** button, the reclassification of the loni surface is executed.

The output from this tool will be stored in *...output/zones/<Current Zone Map>/loni-nbins/grid*, where *<Current Zone Map>* is a reference to whatever value is specified in the **Delineation of Features: Automated Methods>>Current Zone Map** slot when the button is pushed. This intermediate product should not be confused with the final product.

Figure 43 shows an example of the loni bin product. These data are not particularly easy for the human eye to interpret in the cartographic form, although the user should be able to discern shifts in the colorization scheme on a per-subcatchment basis. The identification numbers used to generate the colors are unique across all subcatchments. This global identifier is seen in the column labeled *nac-nmru* in figure 44. For example, global loni bin identification numbers for subcatchment 1 will range from 1 to 30. For subcatchment 2, they will range from 31 to 60. Within each subcatchment, the loni bins are numbered from 1 to the number of loni bins for that subcatchment (that was specified in figure 42). These local loni-bin identifiers are seen in the column labeled *nac* in figure 44. The subcatchment identifiers are shown in the column labeled *nmru*. The actual value of loni

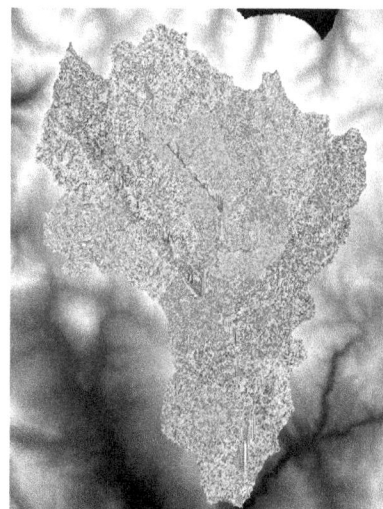

Figure 43. An example of GRID output for the **Topmodel-type Zones>>Loni Bins** tool.

associated with the loni bin is referred to as *st* (fig. 44) and is actually considered a parameter and is discussed in the "Parameter Methodologies" appendix. The *ac* column (fig. 44) also is a parameter. It indicates the percentage of cells in the subcatchment that has this bin's range of loni values.

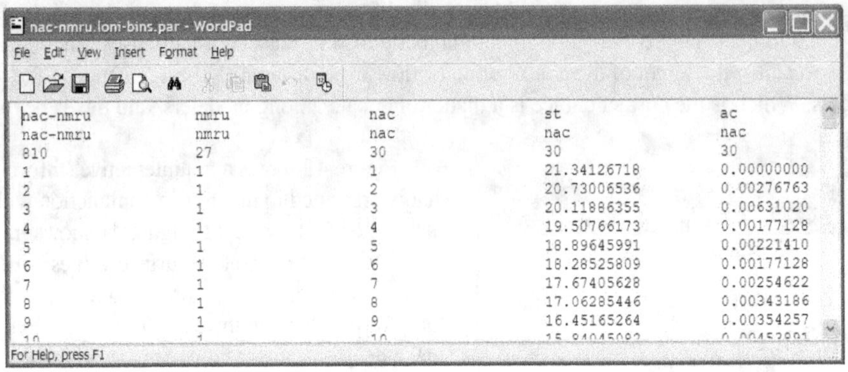

nac-nmru	nmru	nac	st	ac
nac-nmru	nmru	nac	nac	nac
810	27	30	30	30
1	1	1	21.34126718	0.00000000
2	1	2	20.73006536	0.00276763
3	1	3	20.11886355	0.00631020
4	1	4	19.50766173	0.00177128
5	1	5	18.89645991	0.00221410
6	1	6	18.28525809	0.00177128
7	1	7	17.67405628	0.00254622
8	1	8	17.06285446	0.00343186
9	1	9	16.45165264	0.00354257
10	1	10	15.84045082	0.00453891

Figure 44.   An example of ASCII output for the **Topmodel-type Zones>>Loni Bins** tool.

Figure 44 is an example the parameter output that is usually associated with the loni-bin zone map. It is not derived at the time the loni-bin map is delineated, but is shown here to point out that it is generated from a specialized table that is associated with the zone map produced by the **Topmodel-type Zones>>Loni Bins** tool. This specialized table shares the name of the loni-bins grid, but has a ".loni" suffix. The user should take care when managing (for example, copying, renaming) the loni-bin GRID to ensure that the specialized table is propagated to the new GRID.

The output from this tool will be stored in ...*output/zones/<Current Zone Map>/loni-bin/grid*, where *<Current Zone Map>* is a reference to whatever value is specified in the **Delineation of Features: Automated Methods>>Current Zone Map** slot when the button is pushed. The user will be prompted to name a copy of this output that will be stored in the Write Directory.

Other Zones

A variety of miscellaneous tools are grouped here. As there is no unifying basis to this "kitchen sink" grouping, the discussion will proceed directly to the introduction of the individual tools.

Combine Theme w/Current Zone Map

This tool is essentially a tool for the spatial intersection of raster datasets. After invoking it, the user will be prompted to specify the name of a second GRID, via the **Grid Manager**, to intersect with the Current Zone Map. All zones in both of the inputs will persist in the output with unique identifiers (although not necessarily the same numbers as in the input). In addition, all combinations of the two inputs will be uniquely identified, effecting new zones.

This tool is generally useful for merging the output from a GIS Weasel delineation tool with that of another tool, or with an externally created GRID. An example is if a user created a two-plane type map (described in the "Contributing Areas: One, Two, Three" section) and wanted to sub-divide those zones on the basis of vegetation patches. The user could set the **Tool Panel>>Current Zone Map** to either of the GRIDs, invoke **Tool Panel>>Automated Methods>>Combine Theme w/Current Zone Map**, and specify the other GRID using the **Grid Manager**.

The only output from this tool is the one named by the user. No copy is stored in the Output Subdirectories.

Squares (NNNY-NNNX)

This tool is used to generate zone maps that look like the map presented in figure 36. The term *lattice* is really a synonym for grid, but was selected to help differentiate this output from the standard ArcInfo GRID that is so broadly discussed elsewhere in the manual. It should be noted that the cell size of all GRIDs produced by the GIS Weasel, including those produced by the **Squares (NNNY-NNNX) Generation** tool, is consistent with the cell size of the Input Elevation Grid. The **Squares (NNNY-NNNX) Generation** tool will simply make square zones of a user-specified size, built out of GRID cells whose size matches those found in the Input Elevation Grid.

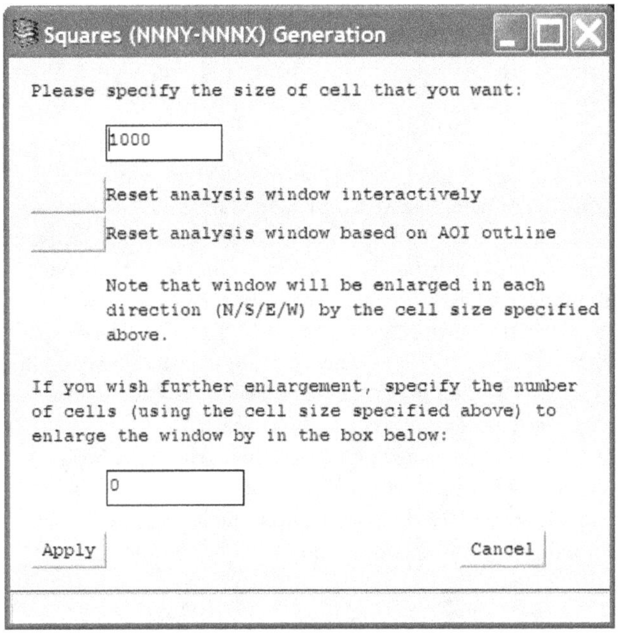

Figure 45.   The **Squares (NNNY-NNNX) Generation** menu.

The **Squares (NNNY-NNNX) Generation** menu (fig. 45), allows the user to specify the length of one side of the square zones to generate in the top-most slot on the menu. The squares will be orthogonally oriented (that is, aligned in columns and rows that parallel the sides of the Analysis Window).

By default, the spatial extent of the map produced will be the Analysis Window, unless specified otherwise using the buttons labeled **Reset analysis window interactively** or **Reset analysis window based on AOI outline**, or typing a value into the slot at the bottom of the menu. The **Reset analysis window interactively** button will allow the user to use the **Pan/Zoom>>View Control** tools to adjust the field of view, which will then be used to redefine the Analysis Window. The **Reset analysis window based on AOI outline** is fairly self-explanatory. The tool will inspect the AOI map, extract the minimum and maximum coordinates that define the boundary, and use this to reset the Analysis Window. This usually results in an Analysis Window that is relatively snug to the AOI. The entry slot at the bottom of the menu allows the user to buffer the Analysis Window in whole zones of the lattice map that is planned.

The output from this tool has been used by a variety of types of model, including groundwater, surface water, and climate models. The only output from this tool is the one named by the user. No copy is stored in the Output Subdirectories.

Elevation Slices

This tool allows the user to dynamically reclassify elevation into bands and use the result to create a standalone zone map. Alternatively, the result can be intersected with the Current Zone Map to make a new zone map. At the top of the menu (fig. 46) presented by the tool are **Statistics** associated with the DEM. The elevation data will be reclassified into some number of bins, indicated by the slot labeled **Num of Zones**. The user can choose between an equal interval (*eqinterval*) or equal area (*eqarea*) reclassification **Method** to derive these bins. The values that define the upper and lower limit for each bin are dynamcally created on the basis of the range of values in the input elevation data, in addition to the other settings specified on the menu.

The equal interval option ensures that each of the bins has the same range between the upper and lower limit as all the other bins. The number of cells in each bin in the output zone map is by no means the same. If this is sought, then the user should select the equal area option. The ranges of the bins that result from this option are not necessarily the same. The numbering of the zones in the GRID output from this tool usually runs from 1 to *n*, where n is the **Num of Zones** value. If the user wishes to count from *m* to *(n+m)*, then *m* should be specified in the **Base Zone Val** slot. If this slot is left blank, *m* is set to 0. If the user wants to exclude elevations above or below some threshold, then that should be specified in the **Min Input Elv** and **Max Input Elv** slots. Further information can be found in the **ArcDoc** under the reference for the SLICE() command. The **Yes/No** radio buttons indicate whether the output should be elevation bands only (by answering *No*) or elevation bands intersected with the Current Zone Map (by answering *Yes*). The **Apply** button will execute the reclassification with the currently specified values. Only the **Method** and the **Num of Zones** values are actually required to run the tool. The **Cancel** button will terminate the execution of the tool without attempting to create any output.

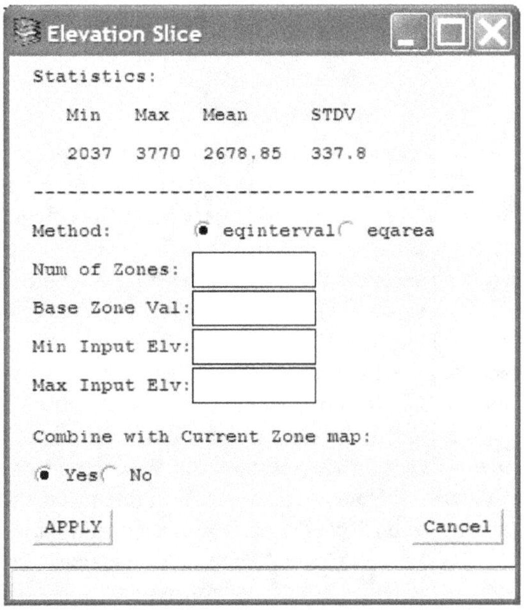

Figure 46.   The **Elevation Slice** menu.

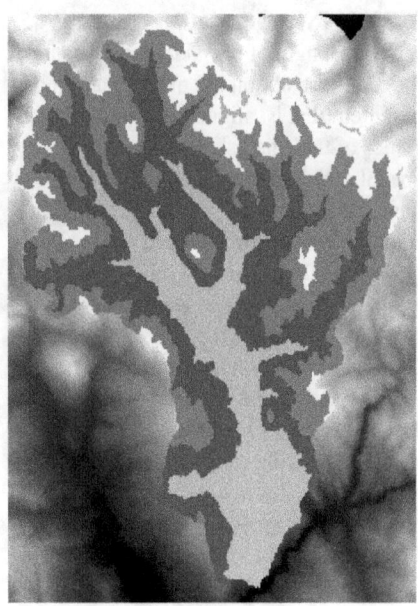

Figure 47. An example of the output from the **Elevation Slice** tool.

Figure 47 shows an example of output from the **Elevation Slice** tool. The only output from this tool is the one named by the user. No copy is stored in the Output Subdirectories.

Zonal Statistics

The Zonal Statistics tool can be used to create an output zone map where the cells in each zone, instead of being marked with an identification number, are marked with a statistic calculated by the tool. Figure 48 shows the **Zonal Statistics** menu. When the user presses the **Apply** button on this menu in the current configuration, the tool will produce a GRID where the cells associated with a zone in the **Zone grid** are all assigned the mean elevation value for that zone. The input **Zone grid** in this example is called *str2knop*. The source of the data for which the statistic is being calculated is specified in the **Value grid** slot (in this case, *dem*). The statistic being called for in the **Zonal function** list is *mean*.

Clicking with the right mouse button in either the **Zone grid** or the **Value grid** slot will invoke the **Grid Manager**, which allows the user to browse the directory tree to find the relevant GRID. The radio buttons under the heading **Handling of NODATA** indicate whether null values should be considered when

calculating the statistic. This is relevant for statistics that use the number of cells in the derivation. If the **DATA** option is selected, then only the cells that have legitimate values within the **Value grid** are counted. An example of how this can affect a calculation is a zone of the **Zone grid** that has 10 cells, but only 5 of the corresponding cells in the **Value grid** have nonnull values. If the *DATA* option has been selected, the mean will be the sum of the 5 values divided by 5. If the *NODATA* option had been chosen, then the sum of the 5 values would have been divided by 10. The **Cancel** button dismisses the tool without creating any output. The **Help** button presents some contextual information.

The only output from this tool is the one named by the user. No copy is stored in the Output Subdirectories.

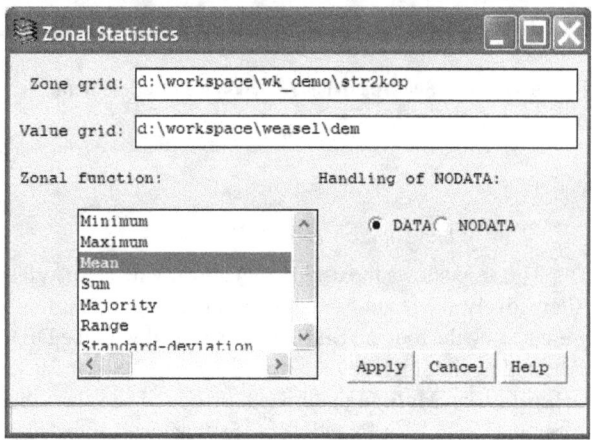

Figure 48. The **Zonal Statistics** menu.

Reclassify a Raster Layer

This tool allows the user to interactively choose a reclassification scheme for a GRID that the user specifies. After specifying a GRID to reclassify with the **Grid Manager**, a **Reclass Control Panel** and the **Reclass Scheme Specification** menu are presented (fig. 49). The **Reclass Scheme Specification** menu is fully editable. Values defining the low and high values of a reclassification bin, as well as the value to assign to the output can be directly typed into the menu. The reclassification scheme can be saved to the Write Directory by pressing the **Reclass Control Panel>>Save** button. Conversely, a pre-existing reclassification scheme can be loaded by pressing the **Reclass Control Panel>>Load** button. A **File Manager** will be presented, and the user must then select a properly formatted reclassification specification. The format followed is that of an ArcInfo *remap* file. For more information on this file format, the user shoud type `remap` into the slot on the **ArcDoc>>Index tab**. The **Reclass Control Panel>>Clear** button will erase all entries from the **Reclass Scheme Specification** menu.

Pressing the **Reclass Control Panel>>Cellvalue** button allows the user to click on the cells of the GRID being reclassified in the main **Map Display** area to determine the values for individual cells. This functions in the same way as the **Tool Panel>>Cell Query** tool. Results are presented to the **Message Board**. The user can terminate the query process by clicking over the **Map Display** window with a nonleft mouse button.

The **Reclass Control Panel>>Auto** button calls the same tool presented by the **Tool Panel>>Automated Methods>>Other Zones>>Elevation Slice** tool. The user is referred to the description of the **Elevation Slice** tool for more information. The **Reclass Control Panel>>Help** button presents information on how to use the menus. The **Reclass Control Panel>>Cancel** button dismisses the tool without attempting to create any output. The **Reclass Control Panel>>Apply** button will carry out the reclassification as specified in the **Reclass Control Panel** and the **Reclass Scheme Specification** menus.

A

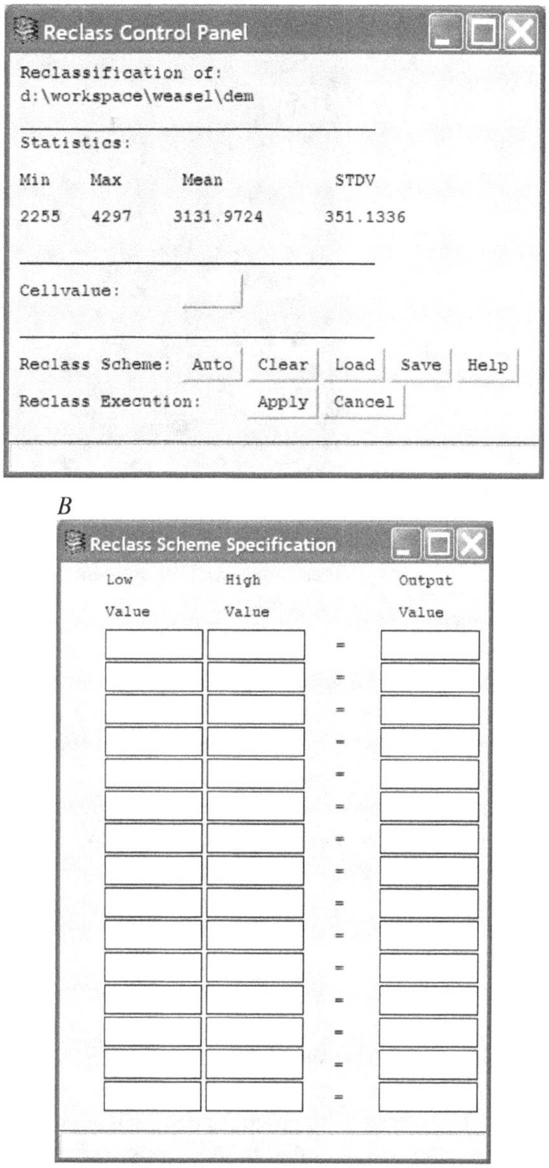
Reclass Control Panel
Reclassification of:
d:\workspace\weasel\dem
Statistics:
Min Max Mean STDV
2255 4297 3131.9724 351.1336
Cellvalue:
Reclass Scheme: Auto Clear Load Save Help
Reclass Execution: Apply Cancel
B
Reclass Scheme Specification
Low Value
High Value
Output Value

Figure 49. *A,* **Reclass Control Panel** and *B,* **Reclass Scheme Specification** menu.

The output from this tool will be stored in *...output/surfaces/<specified GRID>/reclass/grid*, where < *specified GRID* > is a reference to whatever GRID was selected with the **Grid Manager** menu. The user also will be prompted to name a copy of this output that will be stored in the Write Directory.

Split Zones Into Discrete Islands

Figure 50 shows an example of input and output for this tool. The tool is useful if a previous delineation method has produced a zone that is actually composed of multiple islands that are noncontiguous (as resolved by adjacency in both the cardinal and diagonal directions) and the user wishes to uniquely identify each island. The input zone map shown in figure 50*A* was produced by a logical query. All the red cells in this GRID have the same identifier. In the output (fig. 50*B*), each island is assigned a different color (identifier). The only output from this tool is the one named by the user. No copy is stored in the Output Subdirectories.

Outflow Network

This tool will find the single largest outlet (determined using the Flow Accumulation surface) for each zone in the Current Zone Map and use these cells as the starting point for first-order drainage network links. The tool will delineate the drainage network that is initiated at these starting points by using the Flow Direction surface to navigate downhill from these points. Each starting cell will have a unique identifier, and a new identifier will be assigned to each link. See the discussion of the **Tool Panel>>Automated Methods>>Drainage Network** tool for more on how links are defined. This tool is intended to create a drainage network that includes only the links necessary to route streamflow from first-order drainage areas downstream. The expected type of Current Zone Map is a one-plane type of contributing area (described in the "Contributing Areas: One, Two, Three" section).

A B

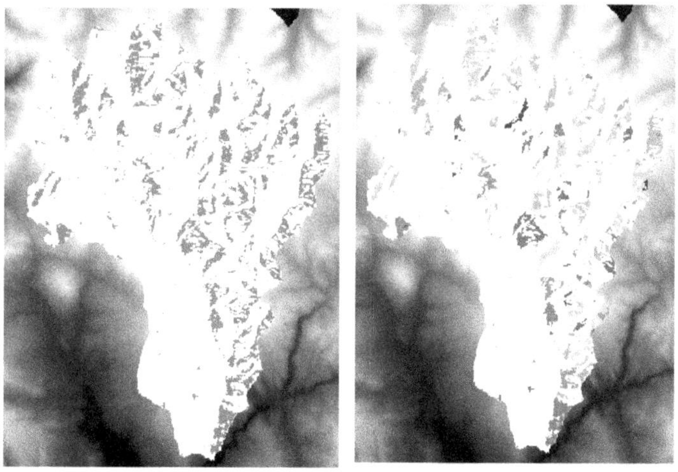

Figure 50. An example of the *A,* input and *B,* output of the **Split Zones into Discrete Islands** tool.

Figure 51 shows an example of the output from the **Drainage Network** tool. The Current Zone Map is represented in the graphic with red lines. These red lines are not part of the output, but are presented to emphasize how the tool functions.

The output from the **Drainage Network** tool will be stored in ...*output/zones/ <Current Zone Map>/route-ordered/grid*, where *<Current Zone Map>* is a reference to whatever value is specified in the **Delineation of Features: Automated Methods>>Current Zone Map** slot when the button is pushed. The user will be prompted to name a copy of this output that will be stored in the Write Directory.

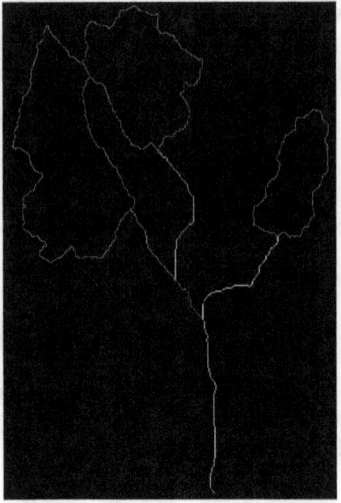

Figure 51.   An example of the output from the **Outflow Network** tool, shown with the input as the red polygons.

Interactive Methods

The **Tool Panel>>Interactive Methods** are separated from the **Tool Panel>>Automated Methods** simply to signify the level of user input for the operation of the tools. The authors hope that this organization is relevant to users.

Id # Manipulation

This tool allows users to change the numerical values associated with the zones in the Current Zone Map. This tool can be used to group multiple zones by assigning them the same value. It also can be used to eliminate zones by assigning the numerical flag for the null value (-9999). The **Id # Manipulation** menu (fig. 52) provides this functionality.

Pressing the **Pick** button allows the user to click on one or more zones with the left mouse button in the main **Map Display** area. Any other button will terminate the selection process. When this happens, a small menu (the **Assign New Value** menu) with an entry slot is presented for specifying the new value to assign to the zone(s). The default value of this menu is the identification number associated with the zone that was first clicked on. When the **Assign New Value>>Apply** button is pressed, the screen will be redrawn so that the modified zone will be colored according to the new identification number. If a zone identifier is set to –9999, it will be seen as white during the use of this tool, but will be eliminated in the output zone map. If any zones were selected in error, the user can press the **Assign New Value>>Cancel** button to discard the currently selected set without changing any identification numbers. If this is done, the tool will still be running through the **Id # Manipulation** menu.

An alternative way to change values is to interact with the **Table of Zone Id values**. The column labelled **Current** lists the values associated with each zone in the Current Zone Map, and the column labeled **New shows** the values that will be associated with the corresponding zones in the output. Initially, these two columns are identical because no changes have been made. Changes to the table will not result in the main **Map Display** area being refreshed. The **Id # Manipulation>>Refresh Display** button can be used to effect this. To help the user understand which colors on the map correspond to specific identification numbers, the button labeled **Click on Zone to check Ids** can be used. The value of the zones clicked on with the left mouse button (in the main **Map Display** area) are shown to the right of the text **Id # Manipulation>>Value**. The **Id # Manipulation>>Cancel** button will dismiss the tool without creating any output. The **Id # Manipulation>>Apply** button will use all the changes made to make a new zone map.

The only output from this tool is the one named by the user. No copy is stored in the Output Subdirectories.

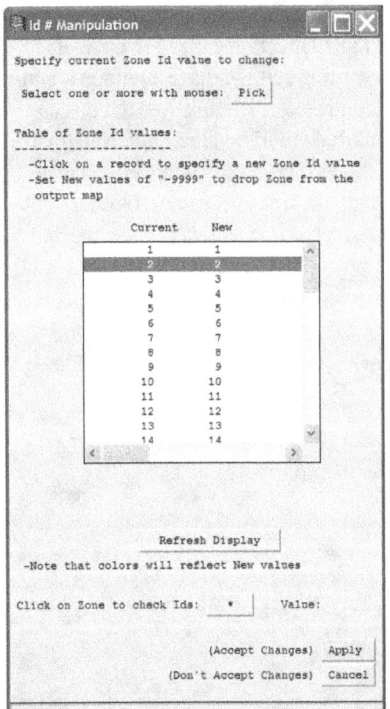

Figure 52.   The **Id # Manipulation** menu.

Digitize

This tool allows the user to create new zones or modify pre-existing ones in the Current Zone Map by using the mouse to perform on-screen (sometimes called "heads-up") digitizing. The **Digitizer** menu is shown in figure 53. The value to be written into the GRID is specified by the **New Id Value** slot. The user can type in an alternate value, or select a pre-existing value off the map after pressing the * button, to the right of the **New Id Value** slot. The default value is one larger than the highest number that exists in the Current Zone Map.

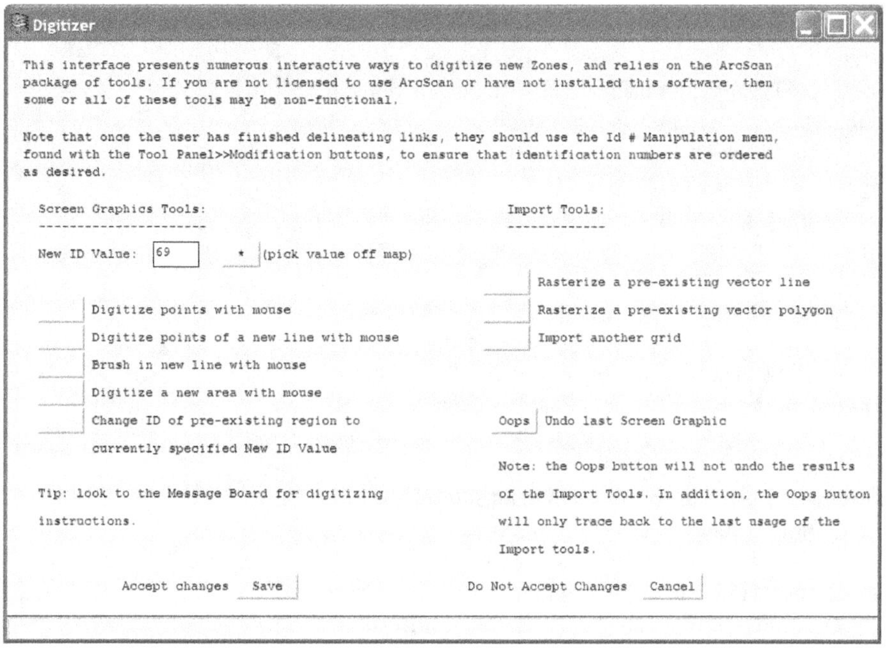

Figure 53. The **Digitizer** menu.

There are a variety of ways to add the **New Id Value** to the screen. The **Digitize points with mouse** button changes the value of a single cell with each click of the left mouse button in the main **Map Display** area. This is useful for when the field of view is very close to the data and individual cells are easily discernable in the **Map Display** area, and only a few cells need to be added.

The **Digitize points of a new line with mouse** button allows the user to click a series of points in the main **Map Display** area with the left mouse button. The **Digizer** will "connect the dots" reassigning not only the points to the **New Id Value**, but all the cells in a straight line path between the points.

The **Brush in new line with mouse** button allows the user to click-and-drag with the left mouse button in the main **Map Display** area. All the cells that the cursor passes over prior to releasing the left mouse button will be assigned the **New Id Value**.

The **Digitize a new area with mouse** button allows the user to click a series of points in the main **Map Display** area with the left mouse button. The **Digitizer** will not only "connect the dots," but will connect the first point with the last to form a closed polygon. All cells on the perimeter and interior of this polygon will be assigned the **New Id Value**.

The **Change Id of pre-existing region to currently specified New Id Value** button operates like the "paint bucket" found in a variety of graphics packages. After pressing this button, the user clicks with the left mouse button in the main **Map Display** area. The **Digitizer** will determine the identity of the cell that was clicked upon and then look for all cells with the same identifier that also are connected to the selected cell. Connectivity is evaluated in eight directions. This tool can be used in a similar fashion to **Tool Panel>>Id # Manipulation** to reassign identification numbers associated with zones.

The **Digitizer>>Import Tools** can be used to bring pre-existing data into the zone map that is being edited. The **Rasterize a pre-existing vector line** tool allows the user to specify a coverage that has lines within it, to select one or more lines ("arcs") from that coverage, and to bring raster versions of those lines into the zone map being edited by the **Digitizer**. The **Rasterize a pre-existing vector polygon** tool functions in the same way, except that the lines are treated as an area (resulting in a "filled in" zone). The **Import another grid** works in the same way; the user can specify the GRID and then select one or more zones from it for integration into the map being edited by the **Digitizer**.

The **Digitizer>>Oops** button will undo all the edits made since the last save event occurred. Using the **Digitizer>>Import Tools** will cause a save event. The **Digitizer>>Cancel** button will dismiss the menu without creating any output. The **Digitizer>>Save** button will use all the changes made to make a new zone map.

An example of how this tool can be used is helpful. Assume a Current Zone Map is a drainage network that has a single link that needs to be broken into two because it is too long. First, the user can use **Digitizer>>Digitize points with mouse** button to change the identification number associated with a single cell on the link at the location where the break should occur. Then the user can use **Digitizer>>Change Id of pre-existing region to currently specified New Id Value** with the same **New Id Value** as used with **Digitizer>>Digitize points with mouse** to click on the link one cell upstream from the single point. This will cause the entire upstream portion of the link to be assigned to the **New Id Value**.

The only output from this tool is the one named by the user. No copy is stored in the GIS Weasel data management system.

Dissolve

It is easy to generate a zone map that has many small zones. In some cases, these zones are legitimate, but it most cases many small zones are spurious. A user may want to eliminate these zones simply because the result will be parameters that describe geographic features that are so small that the influence will never be seen in the aggregate response of the AOI.

The **Tool Panel>>Dissolve** tool will find such spuriously small zones from the Current Zone Map and then arbitrarily reassign those zones to a neighbor. Which zone is selected as the destination should not be important to the user's modeling application. The zones that should be targeted by this tool are akin to "noise in the system." If the identifier that a zone is reassigned to is in fact important to a user's application, then another tool (such as the **Tool Panel>> Id # Manipulation** tool) should be used. The **Tool Panel>>Dissolve Zones** tool could really have been placed under the **Tool Panel>>Clean Up Methods** grouping of buttons.

The **Tool Panel>>Dissolve** tool will collect a minimum allowable area from the user through the **Dissolve Zones** menu (fig. 54). As mentioned, any zones that are smaller than this area are "dissolved." Rather than leaving null values in lieu of these zones, the zones are reassigned to be part of a neighboring zone that exceeds the threshold. Which neighbor is selected is difficult to anticipate and should be considered arbitrary. In theory, it is the neighbor that shares the longest border with the small zone.

As with the interface for the **Tool Panel>>Automated Methods>>Drainage Network** tools, the user is able to specify the threshold by using one of a number of possible units. The actual threshold used will be the **Cell Count**. Values typed in one of the alternate unit slots will be adjusted to be equivalent to the nearest integer value of **Cell Count**. For example, if the user typed a value of 4 into the **Square Miles** slot, then it might be adjusted to 3.91742071043 so that the **Cell Count** would have an integer value.

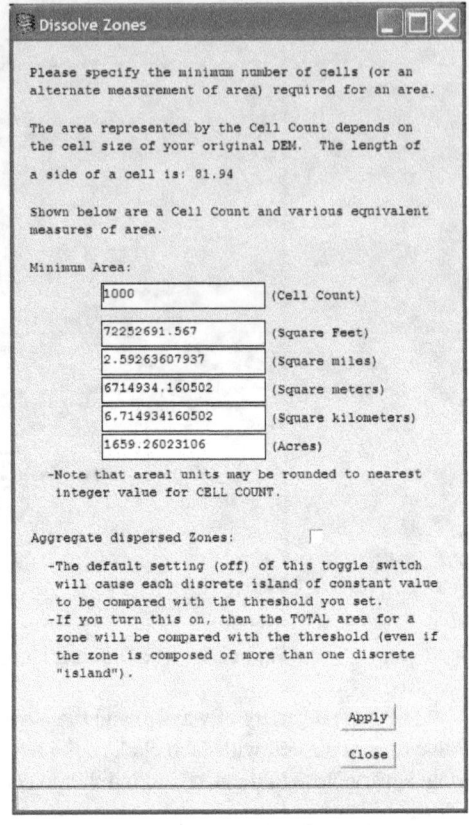

Figure 54. The **Dissolve Zones** menu.

The **Aggregate dispersed Zones** checkbox controls whether the threshold is applied to the entire zone or to each contiguous block or island that constitutes a zone. For most zones that are composed of only a single block, this checkbox will have no effect. For zones that are affected, this can be an important choice. For example, consider a zone made of 2700 cells that is broken into three spatially separated blocks of 900 cells. If the **Cell Count** of 1000 was applied with the **Aggregate Dispersed Zones** checkbox in the *On* position, then the zone would not be affected by the dissolve operation. If the **Aggregate Dispersed Zones** checkbox is in the *Off* position, then the entire zone would be reassigned in the output zone map. This is because the number of cells in each of the islands (900) would be found to fall below the threshold (1000), resulting in the reassignment of each cell of the island to a neighboring zone.

The size of a cell is list in the **Dissolve Zones** menu above the **Cell Count** to help the user understand how much area the **Cell Count** actually represents. Looking at the equivalent value in any of the alternate unit **Threshold** slots also will serve in this. The **Close** button will dismiss this menu without applying the dissolve threshold or deriving a new zone map. The **Apply** button will reassign all zones with areas less than the threshold specified by the **Cell Count** and create a new zone map. If the **Aggregate Dispersed Zones** checkbox is in the *Off* position, then the **Cell Count** will be applied to each spatially distinct block or island of cells.

The only output from this tool is the one named by the user. No copy is stored in the Output Subdirectories.

Logic Query

This tool is the same as the one described under **Tool Panel>>Exploration of Attributes>>Logic Query**. The user is referred to this documentation (see the "Logic Query" section in the "Tool Panel">>"Exploration of Attributes" section). The only output from this tool is the one named by the user. No copy is stored in the Output Subdirectories.

Contrib Area

This tool is basically the same as what is documented in the "AOI Delineation>>Digitize 'pour-point' to create custom watershed AOI" section in the "Setup Phase" section. The user is referred to this documentation. Because this tool is being applied to a pre-existing zone map, it can be used to subdivide or group zones in the Current Zone Map.

Before using this tool, the user is asked *"Do you want to limit the seeding to a set of zones?"* If the answer is *Yes*, then the user selects zones from the Current Zone Map in the main **Map Display** area by clicking on them with the left mouse button. Answering *No* to the query will cause all of the terrain above the selected pour-point (all the way to the topographic divide) to be put into a single, new zone. This can have the effect of grouping all the zones or portions of zones above the point. Answering *Yes* will yield a zone with only the terrain that is both above the pour-point and within the selected zones. The seeding was limited to the white zone in the example shown in figure 55*A*. The location of the pour-point (the black dot) is superimposed on figure 55*B* for emphasis and is not actually part of the new zone map. We can see that the zone delineated above the pour-point does not include the upstream neighbors (the blue, red, and green zones) because of the limit placed on the processing.

The only output from this tool is the one named by the user. No copy is stored in the Output Subdirectories.

Drain from Point

The **Tool Panel>>Drain from Point** tool is intended to provide the user with a way to add a line to a drainage network that is hydrologically correct (according to the DEM). Although the user might be able to create a new line by using the **Tool Panel>>Digitize** tool that looks correct to the casual observer, ensuring that the new line is hydrologically correct can be extremely difficult. Hydrologic "correctness" is defined as the delineation of a new drainage network link that follows the flow direction for every single cell constituting the line. One of the reasons that this is so difficult is that the flow direction is not readily visualized. Rather than forcing the user to specify every cell in the new line, the **Tool Panel>>Drain from Point** tool asks the user to specify only the cell to use as the upstream end (that is, the initiation point) of the line. The tool will identify the cells that are downhill from this point according to the flow direction, thus ensuring the hydrologic correctness of the delineated line.

Once the **Tool Panel>>Drain From Point** button has been pressed, the user is prompted to adjust the field of view within the main **Map Display** area (by using the **Pan/Zoom>>View Control** tools) to allow the user to select the individual cell that will be used as the initiation point for the new line of drainage. When the appropriate magnification has been attained, the **Drain Point Selection>>Okay** button should be pushed. The mouse cursor will then be shown within the main **Map Display** area as a crosshair. The user should move the crosshair over the appropriate cell within the main **Map Display** area and click with the left mouse button. The resultant line will begin at the selected point and end at the downstream cell whose flow direction points off the DEM. An analogy to explain how this tool works is to think of placing a ball on a hill and releasing it. The location at which the ball is released is the initiation point. The path that the ball takes going (all the way) downhill is the new line. The new line does not terminate when it arrives at or flows into a part of the (pre-existing) Current Zone Map. The line will ignore this conflict and proceed to the edge of the DEM.

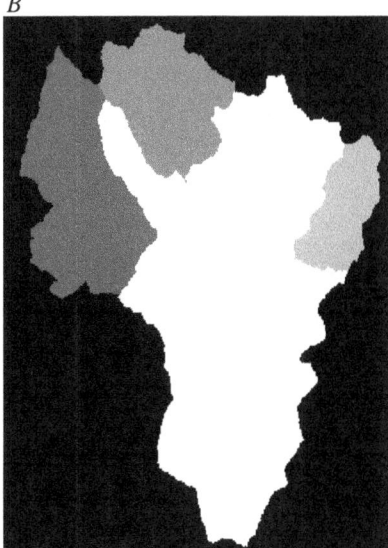

Figure 55. *A*, an example of input to the **Contrib Area** tool. *B*, Output from this tool, using this input and a user-specified point (shown as a black dot).

Once the tool determines all cells downstream from this point, the user is presented with the **Drain Point Overlay** menu, shown below in figure 56. The purpose of this menu is to offer the user of a choice of how the newly derived line of drainage will be integrated into the Current Zone Map. There are three methods available. Figures 57 *B*, *C*, and *D* show the results of these three buttons, all based on the same Current Zone Map (fig. 57*A*, and approximate the same location for the initiation point.

The top button in the **Drain Point Overlay** menu causes the new line to be assigned a single identifying number that supersedes anything already present within the Current Zone Map. The new map will contain the identification of the cells constituting the new line and, where this new line does not exist, the identification of the cells from the Current Zone Map. An example of this output is shown in figure 57*B*. Notice the new line that was added at the lower left of the map. The same color (that is, identification number) is used for the new cells that were not in the drainage network shown in figure 57*A*), as well as all the cells below where the new line merges with the drainage network.

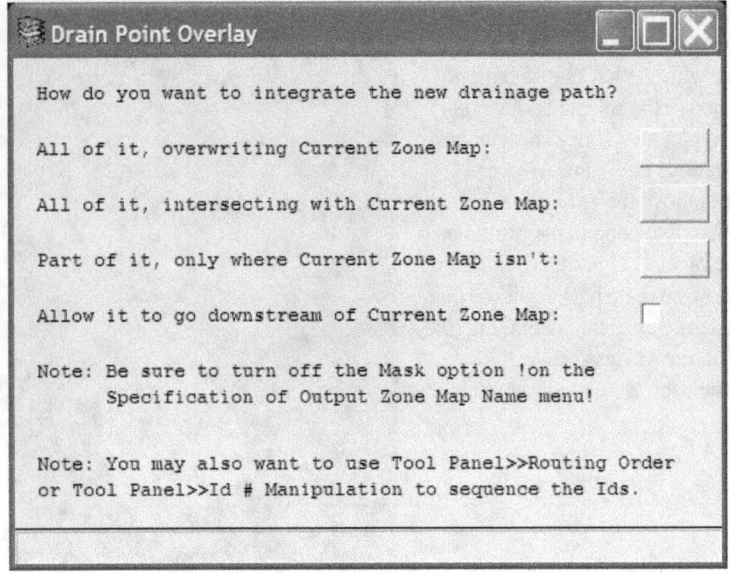

Drain Point Overlay

How do you want to integrate the new drainage path?

All of it, overwriting Current Zone Map:

All of it, intersecting with Current Zone Map:

Part of it, only where Current Zone Map isn't:

Allow it to go downstream of Current Zone Map:

Note: Be sure to turn off the Mask option !on the
      Specification of Output Zone Map Name menu!

Note: You may also want to use Tool Panel>>Routing Order
or Tool Panel>>Id # Manipulation to sequence the Ids.

Figure 56.   The Drain Point Overlay menu.

The middle button on the **Drain Point Overlay** menu invokes a similar option as the top button, except that the cells where both the Current Zone Map and the newly created line exist will be combined to create new zones. The main effect of this is creation of a new zone downstream from the confluence of the new drainage line with a pre-existing zone in the Current Zone Map. This new zone will extend from this confluence point down to the next confluence that existed within the Current Zone Map. An example of this output is shown in figure 57C. Notice that downstream from the confluence of the new line with the yellow line in figure 57A, a new color (signifying the identification number) has been assigned.

The lower button on the **Drain Point Overlay** menu causes the new line to be added where the Current Zone Map does not have any values, but does not disturb the numbering of zones within the Current Zone Map. The new line was added to figure 57D, but no new identification was associated with cells below the point of confluence between the cells in the new line and those in the pre-existing zones of the Current Zone Map.

The only output from this tool is the one named by the user. No copy is stored in the Output Subdirectories.

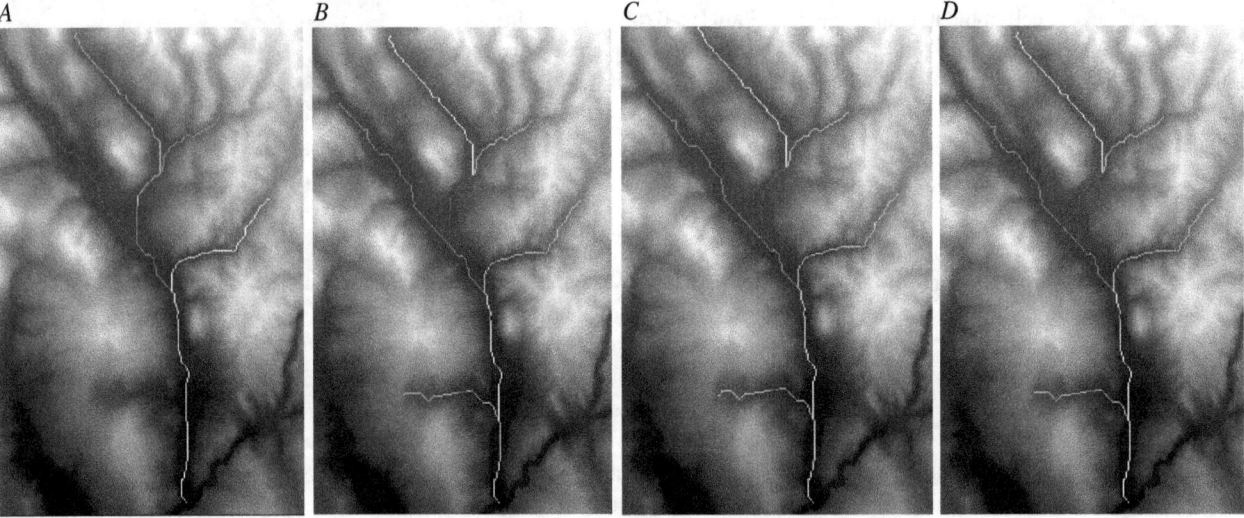

Figure 57.   A, an example of an input to the Drain Point tool and B, C, and D output based on choice made within the Drain Output Overlay menu (fig. 56).

Buffer Streams

Because the areas near to streams are important to so many types of environmental simulation modeling, the GIS Weasel provides a tool to extract these types of zones. Figure 58 shows the **Buffer Streams** menu, which is invoked by pressing the **Tool Panel>>Buffer Streams** button. This tool does not assume that the Current Zone Map is the drainage network to be buffered. Instead, the user must specify this information in the **Buffer Streams>>Drainage Network** slot. Clicking in this slot with the right mouse button will invoke the standard **Grid Manager** tool to help specify the drainage network.

The length of the buffer radius (that is, the width of the buffer to either side of the stream) can be the same for all links within the drainage network, or it can be varied as a function of the stream order of each link in the network. If the *Constant* option is selected at the time the **Buffer Streams>>Apply** button is pressed, the **Buffer Radius** value form is presented to the user.

If the **Stream Order** option is selected from the **Buffer Streams>>Buffer Radius**, then the user will be presented with a secondary question as to the type of stream ordering to use. The options for stream ordering are *Shreve* and *Strahler* ordering. Shreve increases the stream order of the link downstream of a confluence to one higher than the highest order of the incident tributaries. Strahler ordering will increase the stream order of the link downstream from a confluence by one if the two tributaries are of an equal stream order. Generally, the *Shreve* option yields greater differentiation of orders among the links in the drainage networks. The reader is referred to the **ArcDoc** entry for the STREAMORDER command, or any standard hydrology textbook for further information on stream ordering and the meaning.

The user also has the ability to specify the way in which distance from the links in the drainage network is measured. 2-D Euclidean, or "straight-line," distance is easiest to understand, but is not always hydrologically correct. A cell may be very close to a link in the drainage network in Euclidean terms, but may be beyond the topographic divide associated with that link. In terms of the distance that water would travel from that cell to the drainage network link, the cell could be much farther away or not connected at all. Distance can alternatively

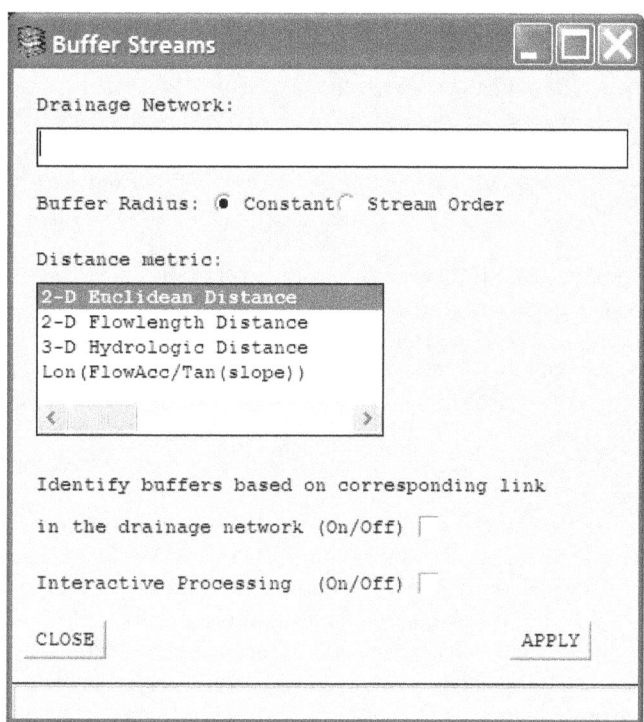

Figure 58. The **Buffer Streams** menu.

be measured according to this hydrologic path over the DEM surface, as specified by the Flow Direction surface. This is normally calculated in the two dimensions that make up the horizontal plane (that is, (x,y) coordinates). This type of distance measure is referred to here as the *2-D Flowlength Distance*. This option is usually the preferred stream buffering distance metric. The appropriateness of this choice ultimately depends on the user's application.

The *3-D Hydrologic Distance* is an attempt to integrate the vertical distance traveled into the calculation of the buffering process. Almost always, this metric is identical to the *2-D Flowlength Distance* and its use is not encouraged because of increased computational costs.

The *Lon(FlowAcc/(Tan(slope))* measure uses the Topographic Wetness Index (referred to as the *loni* value in this manual, described in "The Topographic Wetness Index & Loni Bins" section of "The Delineation Phase>>Automated Methods>> Topmodel-type Zones" section) to buffer the stream. Rather than finding all of the neighbors within some distance of the links in the drainage network, cells whose loni values are sufficiently similar to those found within the drainage network are found. Generally speaking, loni values tend to be relatively low away from streams, high in valley bottoms, and extremely high under the cells that constitute the drainage network. The minimum loni value across all the cells in the drainage network is determined and is used for a threshold. Before applying this threshold, though, it is reduced slightly by subtracting the standard deviation in the loni values across the entire DEM. All cells with loni values equal to or greater than this threshold are identified as being part of the buffer zone(s). A number of cells that are actually spatially noncontiguous with the main grouping around the drainage network links are usually found. Areas that are not contiguous with the main body of cells found by the query process are removed by a subsequent process in the clean-up methods.

There are two remaining settings on the **Buffer Streams** menu. The first specifies **Identify buffers....** The cells that are determined to be part of the buffer surrounding the drainage network are by default all assigned a single identification value, forming a single zone. As a result, the ability to discriminate between parts of the buffer that flow to different links in the drainage network is lost. Further, resolving where (that is, which link) the buffer flows to is lost. The **Identify buffers...** checkbox, when turned *On*, signifies that the cells comprising the buffer should be identified according to which link in the drainage network the cells ultimately flow into.

The final setting on the Buffer Streams menu, the **Interactive Processing** checkbox, asks whether the user would like to be consulted if there are any alternatives to default values for any of the processes involved in the execution of the settings specified by the user. These mostly pertain to the identification of the Flow Direction surface, or in the case of the loni surface creation, the Flow Accumulation surface.

Figure 59 depicts some examples of outputs from the **Buffer Streams** tool on the basis of the drainage network shown in figure 57A. Figure 59A shows a Euclidean distance buffer, where the buffer radius varies as a function of the Shreve stream order. The buffer radii are as follows:

First order = 200 meters
Second order = 400 meters
Third order = 600 meters
Fourth order = 800 meters

The confluence is shown at the upper end of the final link-buffer, shown in yellow at the bottom center of figure 59A. There also is an irregular shape to the upper (which is upstream) end of this buffer on the right side. It looks as if a portion of the buffer has been erased. This is not an error in the processing. The buffer has been generated with the **Identify buffers...** checkbox in the *On* position. Therefore, prior to any buffering, every cell in the entire DEM has been associated with the link in the drainage network to which it drains. Although the missing cells at the upstream end of the final link-buffer are indeed within 800 meters of this link, these cells do not actually flow into that link. These cells flow into the tributary link to the upper right of the final link, although these cells are beyond the buffer distance associated with the stream order of the tributary link (*first order = 200 meters*).

Figure 59B shows the result of using a *Constant* **Buffer Radius**, whose value was set to 800 meters by using the **Distance Metric** on the basis of *2-D Flowlength Distance*. The **Identify buffers...** checkbox was in the *Off* position. This figure shows that although some parts of the network are as broad as the lower link-buffer in figure 59A, there are many areas where the buffer is not nearly so broad. This reflects the distance metric, revealing that flows tend to drain down from valley walls to the valley bottom and then run parallel to the drainage network for a distance prior to merging. Whether or not this is hydrologically accurate is beyond the scope of the current discussion, but the authors want to emphasize the effect of using flow-length types of **Distance Metrics** in the **Buffer Streams** menu.

Figure 59C shows the result for a buffer on the basis of the loni surface, with the **Identify buffers...** checkbox in the *Off* position. Beyond the highly irregular and hydrologically sensitive selection of cells constituting the buffer, the user should note that there appears to be a tributary from the left of the final (most downstream) link-buffer. There was no tributary link in this area of the GRID used for the **Buffer Streams>>Drainage Network** value. The cells in this area have loni values exceeding the automatically determined threshold and form part of a single contiguous zone with cells exceeding the loni threshold in the rest of the network. Although one might argue that this area is not in fact underlain by a link from the drainage network, it also can be said that the analysis has revealed that this area has a high enough loni value to indicate hydrologic character that is substantively similar (and connected) to the drainage network, and should be considered a "riparian" zone. Resolving this question depends entirely on a user's application. If the area is extraneous, then the **Tool Panel>>Digitize** tool can be used to remove it (among other possible techniques).

The only output from this tool is the one named by the user. No copy is stored in the Output Subdirectories.

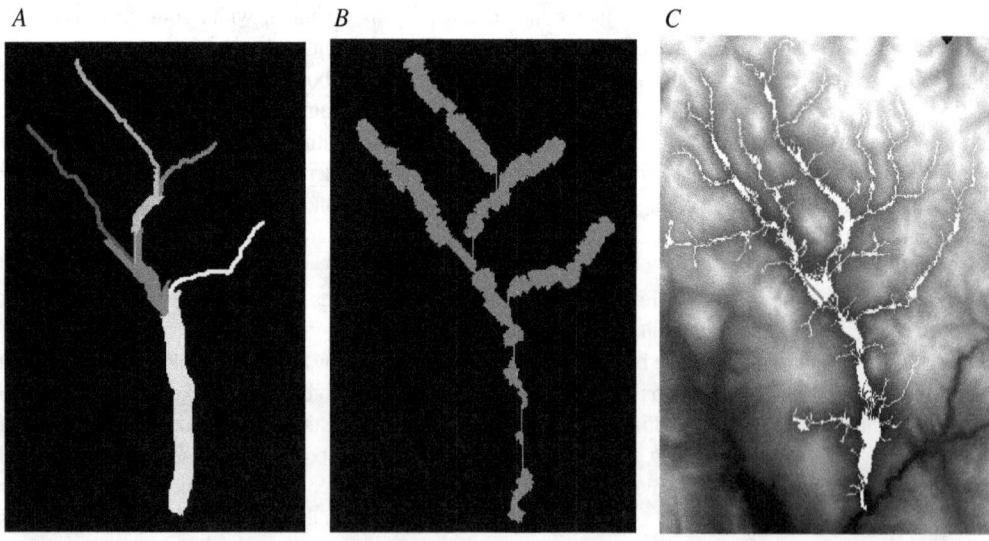

A   B   C

Figure 59.   Examples of output from the **Buffer Streams** tool based on distance measures of A, Euclidean distance B, 2-D Flowlength Distance, and C, Ln(FlowAcc/(Tan(Slope))). All examples use the drainage network shown in figure 57A as the **Buffer Streams>>Drainage Network** value.

Clean-Up Methods

The two clean-up methods are fully automated and could well have been placed onto the **Tool Panel>>Automated Methods** menu. These methods were not because the authors felt that this would obscure the presence of these tools, and that these tools are different from the rest in that these tools are not used for making large modifications to the Current Zone Map. These tools are usually employed as a last step once all the zone shapes have been successfully delineated to adjust how zones are identified.

Pack Ids

This tool is a repackaging of the functionality accessed through the **Output Zone Map>>Pack Identification Numbers...** checkbox, described in the "Overview on Output of Zone Maps" section of the "The Delineation Phase>>Delineation of Features" section. The reader is referred to this documentation. The only output from this tool is a GRID to be named by the user. No copy is stored in the Output Subdirectories.

Routing Order

This tool is used to ensure that the identification numbering associated with the zones of the Current Zone Map always increases in the downstream direction, although the numbers will not always be immediately consecutive when moving to a downstream neighbor. All first-order zones are numbered as a group prior to all higher-order zones. The order in which the first-order zones are numbered is arbitrary. The second-order areas are then located and numbered, and so on until zones of all stream orders have been numbered. The expected input of this tool is some kind of topographically derived zone map, such as a drainage network or a one-plane zone map.

The output from this tool will be stored in ...*output/zones/<Current Zone Map>/nchan-id/grid*, where *<Current Zone Map>* is a reference to whatever value is specified in the **Tool Panel>>Current Zone Map** slot when the button is pushed. The user also will be prompted to name a copy of this output that will be stored in the Write Directory.

The Parameterization Phase

This section provides an overview of the tools accessed through **Tool Panel>> Parameterization of Features**. The purpose of this tool set is to take a set of (GRID) zone maps, derive information about those maps, and create one or more types of ASCII files to contain this information. The ASCII file output is either directly consumed by the user's environmental-simulation model or by some intermediate software that reformats the GIS Weasel ASCII file output into a format that is directly readable by the model.

Overview

Prior to introducing and describing the tools, menus, and other components of the GIS Weasel that are associated with the generation of parameters, several conceptual topics will be addressed. Although the parameterization process is usually almost fully automated, it still requires a small number of inputs from the user. These inputs must be properly specified for the results to be valid for the intended application. The GIS Weasel is a generic tool set and has not been designed to detect errors in these inputs. Therefore, the user should have a firm grasp of the data processing concepts being used by the GIS Weasel prior to parameterization, in addition to those of the modeling application.

Zone Maps Revisited

The GIS Weasel is designed to support the process of generating information about the geographic features that will be used by environmental models to simulate their behavior. This entails the creation of a set of zone maps, where each zone map represents a different type of geographic feature. The zones in a zone map represent instances of the geographic feature type within the AOI. The mechanics of creating zone maps has been discussed in detail in the "Overview on Output of Zone Maps" section of the "The Delineation Phase>>Delineation of Features", but the authors wish to add a few more thoughts on the meaning of zone maps in the context of the data processing that the GIS Weasel carries out in support of environmental-simulation modeling.

Each zone map should contain one and only one type of geographic feature and a zone map should exhaustively depict all instances of that geographic feature type. For instance, a zone map may contain either the links of a drainage network or a set of watersheds, but it should not contain both. If the zone map is used to represent links in a drainage network, *all* links whose behavior will be simulated by the environmental model should be represented. The links of a drainage network should not be spread across two different zone maps. While exceptions to these rules may occur that would be rare.

It also needs to be understood that environmental models do not usually ingest ArcInfo GRIDs, coverages, or any other spatial data file formats directly. The models usually ingest information, referred to here as *parameters*, that describes the features represented in the zone maps. As such, the user should clearly understand the assumptions of both methods by which parameters will be derived and the methods by which the zone maps were created, although the user does not need to understand the details of operating the ArcInfo GIS. For example, a user's model may consume a single value for each instance of some type of geographic feature (that is, for each zone within a zone map) indicating the soil texture associated with that instance. This value is an indicator of central tendency for the soil texture within each zone. The single value is incapable of also expressing or being understood as the heterogeneity of soil texture within each instance. Because of this, the user should take care when delineating the zone map and anticipate this issue. In this example, the creation of zones with very heterogeneous soil texture should probably be avoided, although this is always a function of the user's model.

The Parameter

The term *parameter* is used here to reference any information that is held constant for all the time steps (temporal increments) that constitute the time period over which a model simulation runs. A parameter can describe either geographic or non-geographic entities. The GIS Weasel has been designed to deal with only geographic entities. As a rule of thumb, if an entity can be mapped, then the GIS Weasel can likely be used to generate parameters about it. If a map cannot be made of the entity, then another tool should be used to generate parameters about it.

A parameter is distinguished from the time-varying data that a model ingests by virtue of the fact that a parameter value is constant over the course of a simulation period. An example of this latter type of information, sometimes referred to as *time-series data*, is a set of daily temperature measurements. The GIS Weasel does not provide support for the creation or processing of time-series data, although the GIS Weasel can help if parameters are needed about the locations at which the time-series data are collected (for example, elevations of temperature monitoring stations).

Parameterization Methods and the Parameterization Engine

In support of the generation of parameters, The GIS Weasel provides a library of well over 200 routines. Each routine is referred to as a *parameterization method*. Each parameterization method is documented in the "Parameterization Methods" appendix. Any method can be used to derive information about any zone map. A single method can be applied to more than one zone map. For example, the method for deriving median elevation could be applied to each zone in a drainage network map, as well as to each zone in a map of watersheds.

The shape or semantic meaning of the zones in the map to which a method is being applied, with exceptions noted in the "Parameterization Methods" appendix, is of no importance to the method. The method itself has no understanding that a zone map is supposed to represent the instances of some type of geographic feature within the area of interest, or what the zones inside the zone map (that is, the instances) are supposed to look like. A method will never reject an input zone map as inappropriate or as containing the wrong type of geographic feature.

Each parameterization method will produce a GRID that is automatically stored in the *Output Subdirectories* (described in the "Advanced Topics" section). Each cell in the output GRID can be thought of as being associated with a cell at the same location in the zone map used as input to the method. All the cells in the output GRID that are associated with a zone in the input GRID will carry the value derived for the zone as a whole. For example, the median elevation for each zone (or *link*) in a map of a drainage network is derived, all the cells in the output GRID that spatially correspond to the cells that constitute a given zone (link) will carry the median elevation for the entirety of that zone.

To continue with the example, if link number 1 is composed of 35 cells and the median elevation across those 35 cells is found to be 1,200 meters, then the cells located in the same positions in the output GRID as the ones composing link number 1 in the input zone map of the drainage network will each have the value of 1200. If these data are drawn to a **Map Display** area by using **Pan/Zooom>>Raster Control**, for example, it may be possible to see the same boundaries between links in the output GRID of median elevation that were seen in the original drainage network map. The colors will likely be different in the output than in the original, but the boundaries will be the same. If the **Tool Panel>>Cell Query** tool is used to click on individual cells of the output GRID of median elevation at the locations where link number 1 is found in the input zone map (of the drainage network), the **Message Board** will report 1200 for all these cells.

In addition to a library of parameterization methods, the GIS Weasel provides a set of components that is referred to as the *Parameterization Engine*. These components work together to gather user specifications of which parameterization methods to run against which zone maps, to convert each parameterization method GRID product to an ASCII format, and to provide a cohesive environment in which to apply potentially large combinations of methods. As mentioned, if a parameterization method is run within the Parameterization Engine, then the output GRID will be further processed to reproduce equivalent information in an ASCII format. Strictly speaking, the parameterization method does not carry out the conversion of the GRID output into an ASCII format, rather the Parameterization Engine will use GRID output that each method has stored in the Output Subdirectories to create the ASCII output. Because there is no single menu or other graphical interface that singly represents the Parameterization Engine to the user, it will not be referred to with bold text within this manual.

Although the name of each parameterization method within the GIS Weasel library and the GRID it produces are fixed, the name associated with the corresponding output in the ASCII file output is specified by the user. The Parameterization Engine attaches these user-specifications to the output. The output from the median elevation example previously discussed could be labeled *channel_elev* or *watershed_elev*, for example. The label to be used for the ASCII version of the output from a parameterization method is not actually known by the parameterization method.

This labeling is important because this is how the names that a model uses to refer to different parameters are attached to GIS Weasel-produced information. The GIS Weasel cannot, of course, guarantee the hard-coded knowledge of the terminology used by any particular model, but it will apply user specified values. This (nonspatial) mapping of terminology used by the GIS Weasel to that used by the user's application is referred to here as *indirection*.   In addition to a list of parameterization methods and labels, the Parameterization Engine collects a list of the zone maps to use as input to each method.

Because the Parameterization Engine does not have knowledge of the names used by a simulation model, the responsibility of correctly specifying this information lies with the user. Although this point is being somewhat belabored, errors concerning it usually result in substantially inaccurate output from the GIS Weasel.

Types of Geographic Features and the Dimensionality of Parameter Arrays

A final concept needs to be formally introduced prior to beginning the description of the components of the Parameterization Engine and the parameterization methods that are exposed to the user. As noted in the Zone Maps Revisited section, each zone map should contain all the instances of a *type of geographic feature*. Here the concept of a generic reference for each type of feature map is introduced.

It is common to generate many different parameters to describe several different types of geographic features during a single GIS Weasel processing session. Rather than explicitly associating the full name of the different zone maps with each of the parameterization methods, generic references to each of the zone maps are associated with the methods. Because a zone map represents distinct type of feature, the references can be thought of as placeholders for the specification of the type of feature. This allows the user to perform a one-time assignment of the generic reference to a particular zone map that the user has created within the GIS Weasel (or has available on the hard drive).

This indirection serves a variety of purposes, the first of which is less mouse-clicking on menus for the user. The second is the ability to preconfigure parameterization method lists as well as lists of the types of geographic features in a generic way, before the names of the zone maps are even known. The third is that names assigned to the zones maps created during the delineation phase do not need to match some predefined values. The names of zone maps can be associated with the generic references just prior to parameterization.

In reality, the concept of these generic references comes from several of the environmental-simulation models with which the authors of the GIS Weasel have worked. The arrays of parameters and other structures that the models store information in are often organized according to the type of feature being described. As seen in the example of median elevation for links in the drainage network, most models would store the median elevation for all links in a single array. The size of (that is, the number of values in) the array equals the number of links in the drainage network. The size of the array is referred to as the *dimensionality*. All the other arrays within the model that contain information about the links in the drainage network will have the same dimensionality. To allow the programmers of the environmental-simulation model to reference the size of the array (and do things like loop over the elements in the array) without permanently fixing the array size to a single value, a variable (say, *nlinks*) is used to generically reference the size of the arrays. This variable is sometimes referred to as the *dimension* of a parameter.

As described above, the GIS Weasel attempts to exploit the same concept of indirection that some environmental-simulation models use as a way to be more flexible, convenient, and efficient. Beyond this, the GIS Weasel Parameterization Engine needs uses this information to produce parameter values in the ASCII output that is labeled not only with the parameter names found in the model, but also the names of the dimensions expected by the model. Because the dimension names are not embedded in the parameterization methods, this information also is collected from the user through the components of the Parameterization Engine.

The Components of the Parameterization Engine

After pressing the **Tool Panel>>Parameterization of Features** button, three new menus are presented to the user: (1) the **Feature Type/Method List**, (2) the **Parameterization Control Panel**, and (3) the **Prepared Models** list. Each will be explained in the following section. These menus help the user with two major tasks: the development of a list of what are called *parameter settings*, and the configuration of the execution of the parameter settings.

Developing a List of Parameter Settings

One of the main purposes of these menus is to allow the user to specify the parameterization methods to be executed, the zone map to use as input to each method, and what to name the output within the ASCII output files. Collectively, this information is referred to as the *parameter settings*. The parameter settings re presented to the user in dynamically created menus labeled **Parameter Settings**, as the user makes selections. Prior to making any parameter settings, the **Parameter Settings** menu is not visible.

The Feature Type/Method List

The user may interactively assemble a list of parameter settings by using the **Feature Type/Method List** (fig. 60). This menu allows the user to specify the variable that will be used to reference a geographic type of feature (see the previous "Types of Geographic Features and the Dimensionality of Parameter Arrays" section), and the parameterization method to apply. In the example shown in figure 60, the user is about to add to the list of parameter settings the *param_flowlength.aml* parameterization method. This method is to be applied to the map for the geographic feature type *channels*.

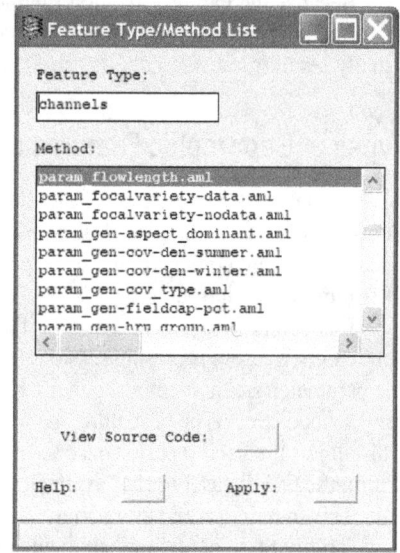

Figure 60. The **Feature Type/Method List.**

By pressing the **Feature Type/Method List>>Apply** button, the user creates the **Parameter Settings** menu, as shown in figure 61. In this example, there was no pre-existing **Parameter Settings** menu. If there had been, then the new information would simply be appended to the pre-existing **Parameter Settings** menu. If the user specifies a large number of parameter settings, then the information may be displayed on two or more **Parameter Settings** menus. This physical separation is because of a limitation in the graphical capabilities of the GIS underlying the GIS Weasel. It serves no logical purpose to the user.

The **Parameter Settings** menu is organized into five columns. Each column is labeled. Each new parameter setting is displayed in a separate row. The ordering of the rows is given by the order in which the user specifies the individual parameter settings. The third and fourth column will contain the **Geographic Feature Type** and **Method**, respectively. Continuing with the example begun in figure 60, the values in the first row of the **Parameter Settings** menu will correspond with the values that were specified by the user in the **Feature Type/Method List** when the **Apply** button on that menu was pushed. The geographic feature type still has not been associated with a zone map. This must be done explicitly using the **Parameterization Control Panel>>Geog Feat Types** button, found in figure 62. The **Parameterization Control Panel** is explained more fully later in this section.

The default **Parameter Name** value, found in the second column of the **Parameter Settings** menu, is *I_need_a_NAME*. The name attached to the ASCII output has no effect on the operation of the parameterization method and is completely up to

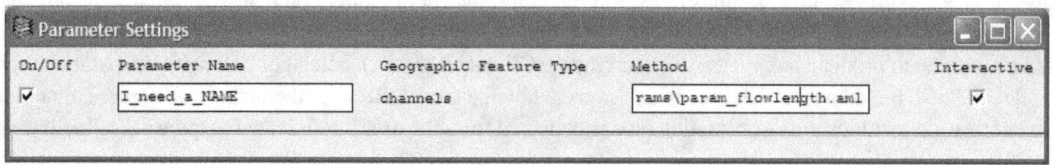

Figure 61. An example of the **Parameter Settings** menu.

Parameterization Control Panel

Geog Feat Types	Clear	Save	Update	Help	Optional Output:

MMS Parameter File: ☑

MMS Input: `c:\workspace\weasel\src_c\mms_param_files`

Is currently specified Data Bin Valid? .TRUE.

MMS Output: `c:\workspace\wk_test\weasel-output.mms`

Data Bin: `c:\workspace\weasel\data_bin`

XML Parameter File: ☐

(Be sure to press <ENTER> after typing)

(TTY)   APPLY   CLOSE

Figure 62.   The **Parameterization Control Panel**.

the user. The user is encouraged to edit this name by typing a new value directly into the **Parameter Name** slot. The same name may be used for more than one output parameter name within the same parameterization effort. The only limitation on the name of a parameter is that it be a single word (contains no spaces).

If the user wishes to remove a parameter setting from the **Parameter Settings** menu (and, therefore, avoid executing the parameterization method), the user can remove the check from the checkbox in the **On/Off** column of the **Parameter Settings** menu (turning the parameterization method *Off*). If the user follows this by pressing the **Parameterization Control Panel>>Update** button (shown in figure 62 and explained later in this section), the row that has been turned *Off* will be dropped from the **Parameter Settings** menu. There is no effect if the **On/Off** checkbox has been turned *Off*, and the **Parameterization Control Panel>>Update** button has not been pressed. In other words, the method will still run. The user can clear the entire Parameter Settings menu by pressing the **Parameterization Control Panel>>Clear** button.

The checkbox at the far right of each row of the **Parameter Settings** menu, under the column labeled **Interactive,** controls whether the user is asked to confirm or change any default specifications used by the parameterization method during execution. Not all methods have user-modifiable defaults within the execution, so even if this checkbox is in the *On* position when the **Parameterization Control Panel>>Apply** button is pressed, the user may not be asked for input. The user is strongly encouraged to place the **Interactive** checkbox in the *On* position during their first usage of any parameterization method to better understand how the method is deriving the output.

If the user has completely specified the parameter settings needed to support their modeling application (or at least, all the ones that the GIS Weasel can satisfy), these settings can be saved for future use by pressing the **Parameterization Control Panel>>Save** button. The user will be prompted to specify a name for the current parameter settings. This name will be added to the bottom of the **Model** list shown in **Prepared Models** menu (fig. 63).

If the user has saved a group of parameter settings, those settings can be recalled by using the **Prepared Models** menu (fig. 63). This can save time and errors associated with having to type in the names of the parameters and geographic feature types. This menu also provides a way for the user to obtain pre-packaged parameter settings for specific models. Clicking on any of the entries in the **Prepared Models>>Model** list will cause a list of parameterization methods, names of geographic feature types, and parameters names to be displayed in the **Parameter Settings** menu. This will not associate the names of geographic feature types with zone maps. This must be done explicitly using the **Parameterization Control Panel>> Geog Feat Type** button, explained later in this section. To use groupings of parameter settings simultaneously, click on each desired entry in the **Prepared Models>>Model** list. When adding a new group of parameter settings to the **Parameter Settings** menu, the user will be prompted as to whether the preexisting values should be cleared. Answering *No* to this question will result in the new parameter settings being appended to the current list.

All entries in the **Prepared Models>>Model** list are configurations that have been developed and distributed by the authors of the GIS Weasel, plus any created by the user. In addition to these, configurations developed by GIS Weasel users are available through the **Prepared Models>>Plug-ins** button. These additional configurations are not documented here (these configurations are documented by the developer). The creation of plug-ins are discussed more fully in the "Advanced Topics" section. As with the **Prepared Models>>Model** list, multiple plug-ins can be selected by pressing the **Prepared Models>>Plugins** button and selecting a plug-in as many times as desired.

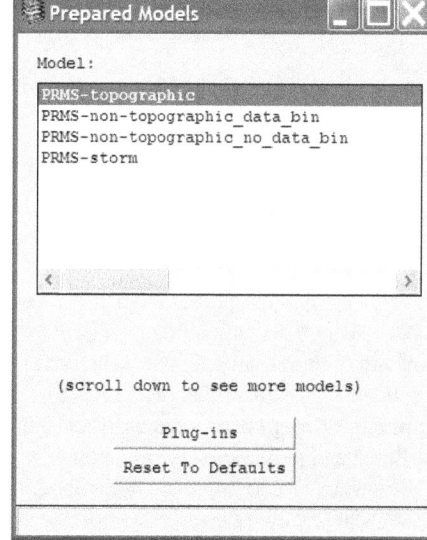

Prepared Models

Model:

PRMS-topographic
PRMS-non-topographic_data_bin
PRMS-non-topographic_no_data_bin
PRMS-storm

(scroll down to see more models)

Plug-ins

Reset To Defaults

Figure 63.   The **Prepared Models** menu.

Figure 64 shows a slightly larger **Parameter Settings** configuration than shown in figure 61, with more than one geographic feature type. This configuration will be used as an example for illustrating the association of the geographic feature type with an actual zone map. It should be noted that the columns that appear in the output file will be sorted according to the order of the rows found here. By pressing the **Parameterization Control Panel>>Geog Feat Type** button (fig. 62), the user will be presented with the **Geographic Feature Type – Zone Map Mappings** menu (fig. 65).

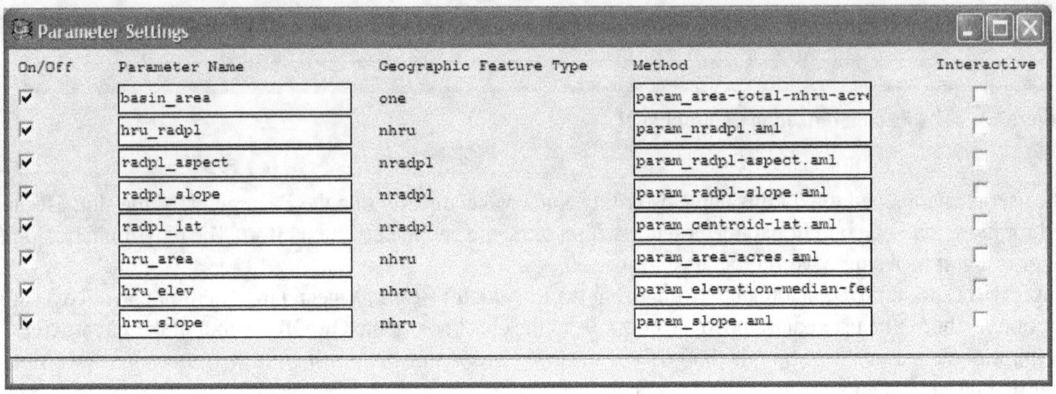

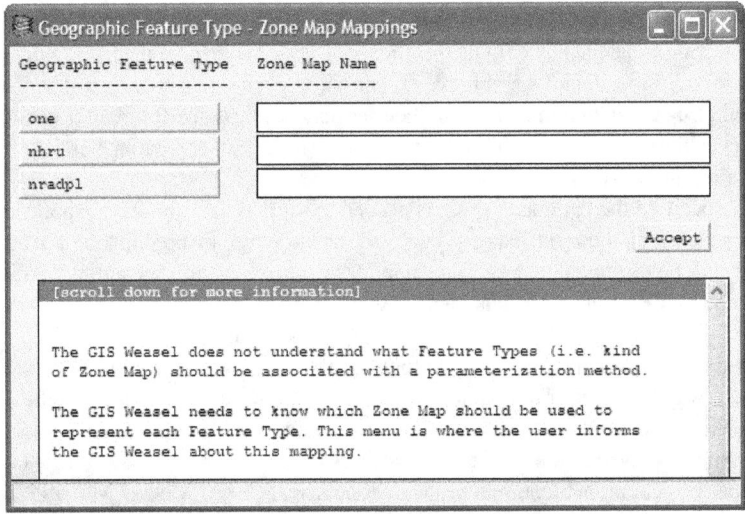

Figure 64.   An example of the **Parameter Settings** menu.

Figure 65.   The **Geographic Feature Type – Zone Map Mappings** menu.

The user may specify the pathname to the zone map to associate with an item in the **Geographic Feature Type – Zone Map Mappings>>Geographic Feature Type** column by typing directly into the slot to the right of each geographic feature type name, or by clicking in the slot with the right mouse button, which will yield a **Grid Manager**, from which a zone map may be selected.

All the mappings on the **Geographic Feature Type – Zone Map Mappings** menu to a specific zone map must be specified before the Parameterization Engine will allow the parameter generation process to begin. If the user presses the **Parameterization Control Panel>>Apply** button before all these mappings have been made, the user will be reminded of the missing mapping. If the user realizes that one or more necessary zone maps are unavailable, the **Geographic Feature Type – Zone Map Mappings>>Accept** button can be pressed and then the **Parameterization Control Panel>>Close** button to return to using the **Tool Panel** and the creation of new maps.

There are several important notes about the zone maps specified with this menu. The zone map simply has to be a GRID that is readable within the user's computing environment. It is not required that the zone map be created by the GIS Weasel. It only needs to have a VAT that contains an integer item, also referred to as a column or field, named VALUE. The VALUE item will be used to identify each instance of the geographic feature type found within that zone map. The zone map, of course, should have the same coordinate system (that is, projection and datum) and at least a partially overlapping geographic extent as the Input Elevation Grid. The zone map can have a different cell size than the Input Elevation Grid, although it should be noted that all GRID products derived from the zone map will have the same cell size as the Input Elevation Grid. In addition, a zone map may be mapped to *more than one* geographic feature type. This can be a convenient way to avoid having to make identical redundant maps.

The **Parameterization Control Panel** menu in figure 62 has already been mentioned to several times. In general, this menu is used to configure the execution of the Parameterization Engine as it executes all of the parameter settings. The **Geog Feat Types** button, as mentioned, allows the user to map the names of geographic feature type to the names of zone maps. The **Clear**, **Save**, and **Update** buttons allow the user to clear, save, and update, respectively, the **Parameter Settings**. The **Help** button provides contextual information about the **Parameterization Control Panel** menu. The **Close** button will dismiss the Parameterization Engine without executing any parameterization methods or producing any output. The **Apply** button will start the Parameterization Engine, causing the parameterization methods to be applied and ASCII output to be generated. The **(TTY)** button allows the user to access the GIS command line. Because this button allows the user to modify the state of the GIS underlying the GIS Weasel and potentially cause run-time errors, it should only be used by advanced users.

The Data Bin

Many of the parameterization methods only require the input of a map of the geographic feature type (that is, the zone map) and maybe a derivative of the Input Elevation Grid. Because all these data sets are well known to the GIS Weasel, these parameterization methods can anticipate the location and names of these data sets. An example of output of this type of method is the area or the median elevation. A large number of other parameterization methods within the GIS Weasel require additional input GRIDs. Examples include those that determine vegetation density or soil texture. These parameters simply cannot be ascertained from a zone map or any of the topographic derivatives of the DEM. Additional information is needed.

To help support the derivation of this parameter type, the developers of the GIS Weasel have assembled several coarse-resolution (1 kilometer cell size) GRIDs into a small set of ArcInfo workspaces collectively referred to as the Data Bin. The **Parameterization Control Panel>>Data Bin** slot allows the user to specify the location of this collection of workspaces. Rather than prompting the user to specify the name and location of each ancillary GRID that might be required by many parameterization methods, many parameterization methods have been developed to look in the Data Bin by using relative pathnames.

The easiest way to determine if a method needs ancillary data is to run it and see if the method asks for it. The user also can refer to the "Parameterization Methods" appendix. Instructions on requesting a subset of the Data Bin, based on a user's AOI, are given in the "Advanced Topics" section. The version of the Data Bin held by the developers of the GIS Weasel extends over only the lower 48 United States.

For users with AOIs that fall outside of the lower 48 United States, parameterization methods that require the Data Bin cannot immediately be used. One option is to assemble replacement GRIDs to populate an imitation of the Data Bin; if all names and formatting matches the standard Data Bin, then the parameterization methods that rely on it will work as normal. Another option is to use different parameterization methods that do not have this dependency. There are a number of parameterization methods whose implementation parallels those that rely on the Data Bin. All these routines are prefixed with *param_gen-*. The *gen* portion of the name refers to *generic*, intending to convey that a Data Bin is not used by the method. These methods will prompt the user during the execution of the Parameterization Engine to specify the locations of the inputs needed. These methods are described in the "Parameterization Methods" appendix.

Once the Data Bin for the AOI has been acquired, the user can specify the pathname to it (including the word *data_bin*) in the **Parameterization Control Panel>>Data Bin** slot, prior to pressing the **Parameterization Control Panel>>Apply** button. The Parameterization Engine will examine one of the GRIDs in the Data Bin and compare a variety of spatial characteristics (for example, coordinate system, geographic extent, numbers of rows and columns, cell size) with those of the Input Elevation Grid. If any differences are detected, the user is notified of these differences in a menu that presents the full description of both GRIDs. In this menu, the user is asked whether to continue using the Data Bin contents despite this difference. If the user responds, *No*, then all parameterization methods that use the Data Bin will not be executed.

The most common difference detected is the cell size. The GRIDs in the standard Data Bin have a cell size of one kilometer. Almost any Input Elevation Grid is going to have a much smaller cell size. This is not generally considered a problem. All derived GRIDs within the execution of the Parameterization Engine will be produced at the cell size of the Input Elevation Grid. Where needed, the values within the coarser cell-sized GRID are dynamically resampled to the appropriate resolution with no notification to the user. A more important problem is if the projection of the Input Elevation Grid does not match that of the GRIDs in the Data Bin. In this case, the user should stop the Parameterization Engine and the GIS Weasel, and fix this problem.

Output

Each parameterization method produces several GRIDs. The logic used in the location and naming of these products is described in detail in the "Advanced Topics" section. The Parameterization Engine controls the execution of the individual parameterization methods and takes these GRIDs to produce simple, columnar ASCII files. The intent of these files is to provide a format that is not only easy to reformat into something directly usable by another kind of software, but also to be easily understood by the user. This provides an excellent way for the user to browse the parameters. An example of this output is shown in figure 66. One such file is produced for each geographic feature type specified in the **Parameter Settings** menu. The files are named according to the convention:

<geographic_feature_type>.<zone_map>.par

where the "<...>" notation is replaced with the name of the geographic feature type and associated zone map. These files are usually referred to as *weasel *.par* files, as distinct from the other optional ASCII outputs that the GIS Weasel can produce. These files are always written to the *.../output/files* subdirectory found beneath the Write Directory.

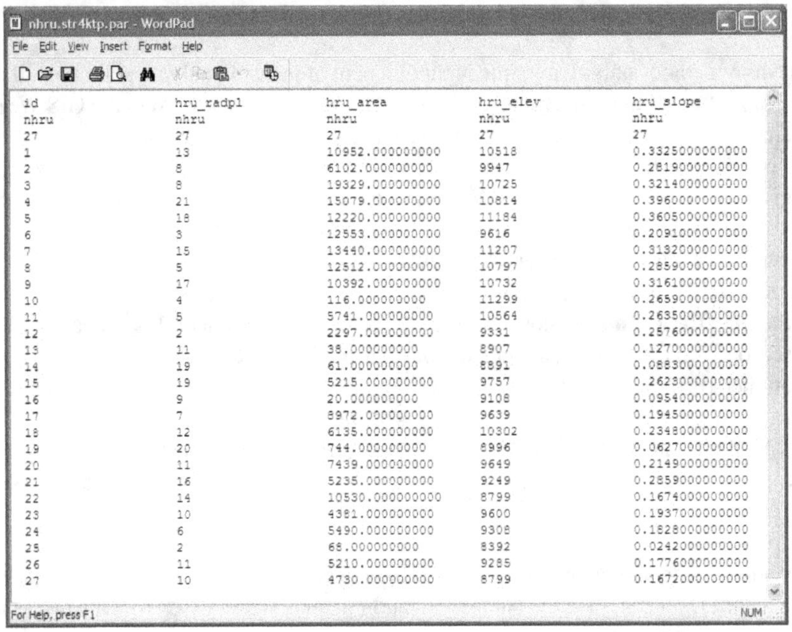

id	hru_radpl	hru_area	hru_elev	hru_slope
nhru	nhru	nhru	nhru	nhru
27	27	27	27	27
1	13	10952.000000000	10518	0.332500000000
2	8	6102.000000000	9947	0.281900000000
3	8	19329.000000000	10725	0.321400000000
4	21	15079.000000000	10814	0.396000000000
5	18	12220.000000000	11184	0.360500000000
6	3	12553.000000000	9616	0.209100000000
7	15	13440.000000000	11207	0.313200000000
8	5	12512.000000000	10797	0.285900000000
9	17	10392.000000000	10732	0.316100000000
10	4	116.000000000	11299	0.265900000000
11	5	5741.000000000	10564	0.263500000000
12	2	2297.000000000	9331	0.257600000000
13	11	38.000000000	8907	0.127000000000
14	19	61.000000000	8891	0.088300000000
15	19	5215.000000000	9757	0.262300000000
16	9	20.000000000	9108	0.095400000000
17	7	8972.000000000	9639	0.194500000000
18	12	6135.000000000	10302	0.234800000000
19	20	744.000000000	8996	0.062700000000
20	11	7439.000000000	9649	0.214900000000
21	16	5235.000000000	9249	0.285900000000
22	14	10530.000000000	8799	0.167400000000
23	10	4381.000000000	9600	0.193700000000
24	6	5490.000000000	9308	0.182800000000
25	2	68.000000000	8392	0.024200000000
26	11	5210.000000000	9285	0.177600000000
27	10	4730.000000000	8799	0.167200000000

Figure 66. An example of the default ASCII format of output from the GIS Weasel.

The first three lines of a weasel *.par file constitute a descriptive header. The first line contains the name of each column. The first column is always dedicated to specifying the identification number associated with each zone in the zone map, regardless of whether this was specified in the

Parameter Settings menu. The entire file will be sorted according to this column. Names in subsequent columns of the first line correspond to the **Parameter Settings>>Parameter Names** that the user specified prior to pressing the **Parameterization Control Panel>>Apply** button. The second line gives the geographic feature type associated with each parameter. The third line gives the number of zones within the zone map used to represent that geographic feature type. The fourth line and below gives the derived parameter values. The columns are separated by spaces.

Optional Output

In addition to the weasel *.par files, the Parameterization Engine can produce two additional ASCII file formats. One is the parameter file format associated with the U.S. Geological Survey Modular Modeling System (MMS) (Leavesley and others, 1996). This text will provide a description of how to create an MMS parameter file by using the GIS Weasel, but does not serve as documentation of that format. The reader is referred to *http://wwwbrr.cr.usgs.gov/mms* for further information on the MMS parameter file format.

The second format is a form of eXtensible Markup Language (XML). XML is a generic encoding scheme. The specific format has been designed to be read by the U.S. Geological Survey Object User Interface (OUI). Again the reader is referred to *http://wwwbrr.cr.usgs.gov/mms* for further information on the OUI parameter file format.

MMS Parameter File

The Parameterization Engine can be instructed to use the parameters from one or more weasel *.par files to overwrite those found in a pre-existing MMS parameter file to create a new MMS parameter file. This is done by turning the **Parameterization Control Panel>>MMS Parameter File** checkbox to the *On* position prior to pressing the **Parameterization Control Panel>>Apply** button.

The reason that a pre-existing MMS parameter file is used in this process is to ensure that the output MMS parameter file will have all the parameter information needed to run the MMS model being used. While the GIS Weasel is capable of producing a large variety of parameters, these parameters describe geographic features only. In many cases, the GIS Weasel will not be able to produce all geographic parameters because of the lack of supporting GIS data. Rather than produce an MMS parameter file that contains something less than the minimum amount of information needed to run a model, the GIS Weasel layers its output on top of pre-existing information. The pre-existing MMS parameter file is specified in the **Parameterization Control Panel>>MMS Input** slot. Clicking in the slot with the right mouse button will yield a **File Manager**, from which a file may be selected.

After all the weasel *.par files have been constructed by the Parameterization Engine, the pre-existing MMS parameter file specified in the **Parameterization Control Panel>>MMS Input** slot is read into memory, then one of the weasel *.par files is read. Wherever a column heading (that is, a parameter name) from the weasel *.par file matches a parameter name in memory, the parameter values in memory are replaced with the values from the weasel *.par file. Where a column heading is not found in memory, it is added to the list of parameters in memory (subject to the constraints described above). When all the column headings have been processed, then the information in memory is written out into an MMS parameter file.

This process actually occurs once for each weasel *.par file. During the first iteration, the pre-existing MMS file used is the filename specified in the **Parameterization Control Panel>>MMS Input** slot and the output is a randomly named file. During the second iteration, the pre-existing MMS parameter file used is the randomly name file produced as output from the first iteration. This continues until all weasel *.par files have been processed. On the last iteration, the output MMS parameter file is named according to the value in the **Parameterization Control Panel>>MMS Output** slot.

If the user wants to generate an MMS parameter file from a set of weasel *.par files without using the GIS Weasel Parameterization Engine, then a conversion routine can be run instead. This can be done by typing *"&run weasel_mms"* within the **Command Line Panel** (press the **Pan/Zoom>>Command Line** button to get this; see figure 6). This converter should be run once for each weasel *.par file by using the output from the previous iteration as the pre-existing MMS parameter file (specified in slot **1** of the menu generated by the weasel_mms program), as described in the previous paragraph.

The process of layering new GIS Weasel-generated parameter information onto pre-existing MMS parameter files can be a convenient device for iteratively developing a set of parameters for modeling. At first, a user can opt for a relatively simplistic set of delineations of geographic features to begin the model calibration process. Initial calibration efforts might focus on non-spatial parameters (for example, related to climate or seasons). Once this information has been set to the user's satisfaction, the user can return to the GIS Weasel to derive more complex maps of geographic feature types and the associated parameters. This new information can be layered on top of the recently calibrated information, replacing only the spatial information that the user has changed and preserving all other information.

Several caveats regarding the generation of a new MMS parameter file are warranted. Some of the software used in this process is written in the C programming language. The C programs are pre-compiled for Windows users, and dynamically compiled for Unix users. In almost all cases, this software runs without issue. If it becomes apparent that there is a problem with this software, the user can choose to recompile this software by first issuing the *make clean* command and then the *make* command from within the *.../weasel/src_c* directory, if these programs are available on their computers. There are a variety of other possibilities by which this software can be recompiled, as well.

Even if the programs are properly compiled, problems can arise because of the MMS parameter file that is used as input to the process. To begin with, this file should be formatted in a Unix style. That is, it should not have "^M" carriage return characters at the end of each line, as DOS-formatted files do. Assuming that pre-existing **MMS Input** file is Unix-formatted and properly follows the conventions of the MMS parameter file format, then the most common sources of problems have to do with trying to merge content of a weasel *.par file that is inconsistent with that given in the MMS Input file. This can occur because MMS is a modeling *framework*, and although it uses a constant file format, the parameters and geographic feature types (that is, dimensions) associated with the specific model to be run within MMS can differ substantially from model to model. The user should make sure that the pre-existing MMS parameter file used as input to this process is indeed consistent with the model that is to be used and the parameter and geographic feature type names used within the GIS Weasel are consistent with the model. Geographic feature types are given within an MMS parameter file as part of the upper portion of the file labeled as *Dimensions*.

If a weasel *.par file specifies the name of a geographic feature type that is not found within the MMS Input file, then the Parameterization Engine will, by design, notify the user by displaying an error message. The rationale for this is that if the **MMS Input** file is so different from what is being described in the weasel *.par file, then it is likely that either the **MMS Input** file is inconsistent with the model being used or the weasel *.par file is inconsistent. In either case, the Parameterization Engine terminates. If the weasel *.par file gives a parameter that is unknown, but that describes a geographic feature type that is in fact present in the pre-existing MMS-format parameter file, then the new information will be added to the output MMS parameter file. If the weasel *.par file gives a parameter that is known, but uses it to describe a different geographic feature type than what is specified in the pre-existing MMS parameter file, then an error will result and no output MMS parameter file will be created.

XML Parameter File

The Parameterization Engine can be instructed to use the parameters from one or more weasel *.par files to create an XML formatted file. This is done by turning the **Parameterization Control Panel>>XML Parameter File** checkbox to the *On* position prior to pressing the **Parameterization Control Panel>>Apply** button. The output will be written to a file called *param. xml* that can be found in the Write Directory.

The format of this file will be described briefly here. The reader is assumed to have an understanding of XML. The root tag enclosing all the information in the file is *<parameterset>*. There are four main tags below this: (1) *<metadata>*, (2) *<dimension>*, (3) *<param>*, and (4) *<value>*. The *<metadata>* tag encloses the specification of the user-id that created the file, the location where it was originally created, the identity of the computer on which it was created, and the date of creation. The *<dimesion>* tag is equivalent to the geographic feature type. This tag is attributed with the name of the geographic feature type, the name of the zone map that it was mapped to, and the number of zones (that is, instances or features) within that zone map. The *<param>* tag defines a parameter, giving the *name*, the geographic feature type that it describes (in the *dim* attribute), the numerical *type* (*float* or *integer*), the *level* of measurement, and the *units* of measurement. The level of measurement can be *nominal, ordinal, interval,* or *bounded*. The first three types are standard and will not be defined here. The bounded type refers to a parameter whose value is actually of the identity of another feature. An example is a parameter named *down_link* that indicates the identity of the drainage network link that is downstream from a feature. The range of values for *down_link* are constrained to the set of identifiers associated with the zone map representing the drainage network geographic feature type.

Sample Usage of the GIS Weasel

This section is intended to serve as a start-to-finish description of the use of the GIS Weasel in support of a specific modeling application. While the details of the delineation and parameterization phases for the application of the GIS Weasel will vary depending on the model being supported, or even with the application of a model to a new geographic location, the user will likely carry out a somewhat similar series of steps.

This section is not intended to serve as an explicit recipe for using the GIS Weasel for all models. No such recipe can be specified. Nor is this section intended to serve as an exhaustive reference of the individual tools found within the GIS Weasel. The preceding sections (and the appendix that follows) provide these details. It is assumed that these sections have been read prior to this section. Further, this section is not intended to serve as documentation for the environmental model that is used as an example here, or even an exhaustive discussion of how to use the GIS Weasel in support of this model. This section is intended to give the reader a sense of the general sequence of events associated with the use of the GIS Weasel.

This section will provide conceptual instructions related to a specific environmental model, procedural instructions for operating the GIS Weasel in support of that model, and discussions of results derived by applying these instructions to the sample data that are provided with the GIS Weasel software package.

Description of the Modeling Application

As has been emphasized a number of times in this manual, the GIS Weasel is a generic tool that has no prior knowledge of specific environmental simulation models. The user is responsible for having expertise regarding the environmental model for which the GIS Weasel will provide inputs. This section gives a brief description of the model being supported in this sample application.

The model is called the Precipitation-Runoff Modeling System (PRMS) (Leavesley and others, 1983). It is a watershed hydrology model that is usually applied to watersheds of 10,000 square kilometers or less. For this example, the model will have a time step of one day (that is, it runs in the "daily" mode). The watershed should not be so large that the time it takes runoff to travel to the watershed outlet exceeds an entire time-step (of one day).

The model requires descriptions of five different types of geographic features that the GIS Weasel is likely to be able to create. These geographic feature types are the basin outline, hydrologic response units, radiation planes, subsurface reservoirs, and ground-water reservoirs. These are referred to within PRMS as `one`, `nhru`, `nradpl`, `nssr`, and `ngwres`, respectively. This section of the manual shows the user how to create a GRID representation (that is, a zone map) of each of these geographic feature types.

The basin outline is usually delineated as the contributing area above an arbitrarily selected point. This point is usually colocated where a stream gage is found, in order to allow the comparison of the streamflow simulated for the basin by the model with the streamflow observed at the gage. Parameters can be adjusted (that is, calibrated) so that the simulated stream-flow matches the observed more perfectly. The contributing area to the point is determined by using the Flow Direction surface derived from the Input Elevation Grid that was specified during the setup phase. There is only one basin outline per application of PRMS.

Hydrologic response units (HRUs) are the spatial units with which the model is most concerned. This type of geographic feature is generally thought of as a land-surface unit that includes both the vegetation above the surface and the soils near the surface. The shape of an HRUs is normally described as a "hillslope," implying a single contiguous area that extends from one bank of a stream segment up to the drainage divide. Two such units are usually associated with each stream segment, one for each bank. A stream segment is usually defined as a link in the drainage network (described in "The Delineation Phase" section). Although streams are not described to the model with parameters, streams are commonly used as the basis from which to derive HRUs. There are many methodologies for delineating HRUs; there is no single "right" method. The modeler/GIS Weasel user must make decisions about which method is the most appropriate for their application.

Rather than treating the entire watershed as a single behavioral "lump," different HRUs can be defined to isolate areas with different hydrologic responses. HRUs enable what is commonly referred to as *distributed* modeling. From another perspective, HRUs can be thought of as aggregations of cells in a GRID where common characteristics (that are described by parameters) are found. While PRMS does not conceptualize hydrological processes occurring within a single spatially lumped entity (that is, a watershed), neither does it conceive of them as occurring in every single GRID cell within a watershed, which would be the most fully spatially distributed manner. Therefore, PRMS should probably be considered a *semi-distributed* model.

Ideally, a single HRU is intended to be homogenous with respect to all the characteristics (described to the PRMS model by *parameters*) that modify the hydrologic response of an area. The larger an HRU, the greater the likelihood that there will be heterogeneity in these characteristics. This heterogeneity is undesirable because each parameter of an HRU is an indicator or measure of the central tendency for some characteristic. A parameter has no way to indicate to the PRMS model of any variation from the central tendency of the characteristic within a given HRU.

A common approach to counteracting the problem of heterogeneity is to delineate smaller HRUs. The user is cautioned that there are practical limits to this approach. In reality, heterogeneity of hydrological characteristics is bound to exist no matter how small the HRUs. Operationally, a large number of HRUs take more computing time to simulate. Statistically, more HRUs increase the number of degrees of freedom in the parameter set, and, therefore, increase the uncertainty associated with the model output.

The user should anticipate a number of factors in determining how much heterogeneity is acceptable in their HRUs and how big or small an HRU should be. The ultimate factor is the sensitivity of the model to heterogeneity. If a user relies on characteristics that are NOT used to generate parameters for the model as the basis for delineating certain spatial features, careful consideration should be given as to why this is being done. If the modeler determines that such a characteristic is an important factor in the behavior of a physical process, then the model structure itself fails to conceptually represent a relevant process. This type of shortcoming can be compensated for through operational tactics, but this requires much more expertise on the part of the modeler.

Even if all relevant processes are simulated by a model, there is often insufficient data available to assess the correctness of simulations on the basis of finer HRUs. For instance, most watershed hydrology models are calibrated by comparing the aggregation of the runoff for all HRUs with the streamflow observed at a gage located at the outlet of the watershed. By comparing the total response of a distributed model with a time-series of data from a single spatial location, the model results are essentially lumped. As long as the aggregate hydrologic response of all the HRUs (and possibly those of the other geographic feature types) sum to the same volume of water observed by the stream gage, then the calibration will show no error. It is entirely possible to get "the right answer for the wrong reason." As a simple example, it is impossible to uniquely identify the correct values for `a` and `b`, if all we know is that `a + b = 5`. Given the inability to decompose which portion of the flow observed at a stream gage came from which HRU, the user should not necessarily use the smallest HRU possible. Doing so would be wasteful in terms of the consumption of disk space and computing (and operator) time. In addition, it would imply a spatially explicit certainty that cannot be validated.

The specific methodology for delineating HRUs should be selected not only on the basis of the concepts represented in the model and concerns relating to validation of the model results, but also as a function of which of the many physical processes typically simulated by a model are important (or dominating) in the particular place where it is being applied. As an example, the hillslope-type of HRU that has been described here is expected to differentiate land surfaces with different aspects (orientations). This separation is especially relevant where PRMS is being applied to watersheds where hydrology is dominated by snow-melt processes. In such watersheds, the amount of solar radiation that a land surface receives is critical to determining the temperatures found on that surface, and hence the amount of snow melt that will occur. If PRMS is being applied in a watershed where no snow accumulates or where the relief is extremely shallow (for example, coastal Florida), this approach may not be as important. It may not necessarily reduce the accuracy of the results, but it is not adding to it either.

Radiation planes also are used by the model to help calculate snow melt. This type of geographic feature is actually derived from the HRU map. Because of this dependency, radiation planes must be delineated after the HRU map has been finalized. The zone map of radiation planes is slightly unusual because it is really a categorical map, like a land-use/land-over map. Each spatially contiguous instance is not identified as a unique object. Instead, each HRU zone is assigned a new identifier that signifies a combination of characteristics (slope and aspect categories). There might be multiple occurrences of the same identifier (that is, combination of slope and aspect characteristics) across a radiation plane zone map.

In addition, PRMS insists that the radiation plane numbered "1" be perfectly flat. This first radiation plane is used by the model as a datum from which to calculate the radiation budgets for all other radiation planes. Because such a radiation plane is by no means guaranteed to actually exist within the watershed being modeled, the GIS Weasel tool for deriving radiation planes will insert a single cell zone at the center of the watershed and assign it the first radiation plane identifier and characteristics.

The two remaining types of geographic features are subsurface reservoirs and ground-water reservoirs. There is not a wealth of GIS data available to delineate maps of these underground features. As a result, users frequently opt to reuse the GRID that represents the basin as the depiction of these feature types. This has the advantage of reducing the number of parameters that PRMS handles, and therefore reduces both the degrees of freedom associated with these inputs and the uncertainty associated with the output. The disadvantage is that the modeler has fewer options for controlling the speed and duration ("flashiness," attenuation) of the hydrologic response of the watershed. Another commonly chosen alternative is to reuse the GRID that represents the HRUs as representations of the subsurface and ground-water reservoirs. This provides the modeler with a large number of parameters for controlling the segmentation of flow out of the basin into base flow and interflow, in addition to the surface runoff generated by the HRUs.

Processing Overview

Before discussing technical details, it is helpful to outline how the recently discussed concepts associated with the PRMS watershed model correspond to a broadly described sequence of steps within the GIS Weasel. After amassing an appropriate GRID of elevation to use as input, the user will start the GIS Weasel and specify the name of this GRID. After creating some derivative GRID surfaces (for example, flow direction or flow accumulation) from the DEM, the user will define the AOI. The AOI will eventually be used, during the parameterization phase, to derive parameters that describe the entire basin to PRMS. The AOI delineation step will be the last point for user-input during the setup phase.

At the beginning of the delineation phase, the user will define a drainage network from the DEM. The resultant network should be geographically constrained (that is, masked, or "cookie-cut") to those areas found within the AOI. The drainage network actually will not be used to derive parameters, but will be used to derive the map of HRUs. Many parameters will be generated by using the HRU map during the parameterization phase. Once the HRUs have been created, then the GRID of radiation planes can be created. This will conclude the delineation phase.

When the user has maps representing the entire basin, the HRUs, and the radiation planes, the parameterization phase begins. First, the user will start the Parameterization Engine and configure the list of parameters to be derived from the three maps (GRIDs of the AOI, HRUs, and radiation planes). Once this list is configured, but prior to having the Parameterization Engine derive this information, the user will specify several additional pieces of information. This will include associating the generic names for the different geographic feature types with the names of the three user-created GRIDs of the AOI, HRUs, and the radiation planes. The user also will specify the location of the data bin and notify the Parameterization Engine that the standard GIS Weasel *.par files should be reformatted into an MMS-format parameter file. Once this information has been specified, the user tells the Parameterization Engine to derive the parameters, and both GIS Weasel *.par files and an MMS-format parameter file are generated. Once this is done, the user will have completed the GIS Weasel processing session.

Preparation for a GIS Weasel Processing Session

As mentioned in "The Setup Phase" section, the only required input to start the GIS Weasel is an ArcInfo GRID of elevation data (that is, the *Input Elevation Grid* or *DEM*). The DEM can have any projected coordinate system. The use of geographic coordinates (that is, latitude and longitude) is discouraged. Also noted earlier, the user should determine the coarsest acceptable cell size for the DEM. A rule of thumb is to ensure that there are between 50,000 and 100,000 cells within the AOI. While having more cells is not likely to improve the performance of PRMS, it will substantially increase the cost of operating the GIS Weasel.

The quality of the DEM directly affects the quality of all the topographic derivatives. The methodology described below for PRMS will rely heavily on the DEM. If the quality of the DEM is poor, then the user should evaluate whether it is poor enough to negatively affect the accuracy of the results produced by PRMS. This is an important point. Even if the HRU zone map that is produced is visually unappealing or even inconsistent with reality, the model may not produce substantially poorer results. If it is decided that the simulation will in fact be negatively effected to a substantive degree, then the DEM should be modified prior to starting the GIS Weasel. There are a variety of techniques available, including "burning" the paths of vector streams into the DEM, "building walls" where the drainage divides should occur, and applying high and/or low frequency pass filters to the data. The user may even choose to interpolate an entirely new DEM from point, line, and contour data by using a package like ANUDEM (Hutchinson and others, 2001) or TOPOGRID (ESRI Inc., 2001)

For most hydrological modeling applications like PRMS, the streamflow simulated for the watershed is compared to the streamflow observed at a stream gage in order to calibrate the model. Because of this, most PRMS-based applications of the GIS Weasel define the AOI as a watershed or contributing area terminating at the location where the gage against which the model will be calibrated is found. Therefore, the user should have a very good idea of where the gage is located on the DEM. In fact, the most reliable way to find this location is to create an ArcInfo coverage of the stream gage location. Although more than one point can be stored in this coverage, only one is used for creating a contributing AOI above that point. The coordinate system of this coverage should be consistent with that of the Input Elevation Grid. The GIS Weasel does not provide tools for the creation of such a coverage, nor will it dynamically reproject the coverage into the coordinate system of the DEM.

The user also should gather any ancillary GIS data needed to make decisions about the delineation and parameterization of features. This might include data such as coverages of roads, rivers, land use, or anything else that the user feels is pertinent to the model application to the AOI. A point coverage showing the locations of the temperature and precipitation measurement stations that will be used to drive the PRMS model is often useful to users. The data bin, described in "The Parameterization Phase" section and the "Advanced Topics" section, is worth special mention. Because the data bin is not required for delineation, the user is free to assemble it at any time prior to parameterization. The user should understand that PRMS derives 10 parameters, relating to soil and vegetation characteristics, from the various layers in the data bin. If no data bin exists, then the user should make alternate plans for providing values for these parameters to the PRMS model.

The Setup Phase

Once the GIS Weasel has been started, the user should specify the **Write Directory** and the **Input Elevation Grid** on the **Write Directory and Input Elevation Grid Specification** menu (fig. 4). For the purposes of this section, the Write Directory will be set to D:\Workspace\wk_demo. The user is prompted to choose whether creation of a new Write Directory is wanted. If this does not occur, then either the Write Directory already exists, or the user failed to press the <enter> key after typing in the **Write Directory** slot. If the Write Directory specified already exists, choose a new name (remember that this section will still refer to D:\Workspace\wk_demo). The Input Elevation Grid will be D:\Workspace\weasel\dem.

Leave the checkbox labeled **Use Input DEM previously associated with current Write Directory** in the default position (*Off*). This is a new Write Directory, so there is no previously used DEM. Leave the **Interactive Processing (On/Off)** checkbox in the default position (*On*), so that you can see all the processing described in this section. No plug-ins will be used here, so the **PLUG-INs** button should not be pressed. Press the **Apply** button.

Answer *No* to the question of whether to set a limit on the Analysis Window. The full extent of the DEM will be used as a result of this response. Answer *Yes* to the question of whether to fill the depressions in the DEM. As a result of this, D:\Workspace\wk_demo\output\dem\filled\grid will be created. Answer *Yes* to the question of whether to derive the Flow Direction surface. As a result of this, D:\Workspace\wk_demo\output\dem\filled\flow-direction\grid will be created. Answer *Yes* to the question of whether to derive the Flow Accumulation surface. As a result of this, D:\Workspace\wk_demo\output\dem\filled\flow-direction\flow-accumulation\grid will be created.

The user is now presented with the **AOI Delineation** menu (fig. 18) and the **Map Display** window, which shows the DEM. An AOI will be created by determining the contributing area to a point where there are stream-gage observations. Prior to starting the **AOI Delineation>>Digitize 'pour-point'**… tool, superimpose a point coverage onto the map to determine the

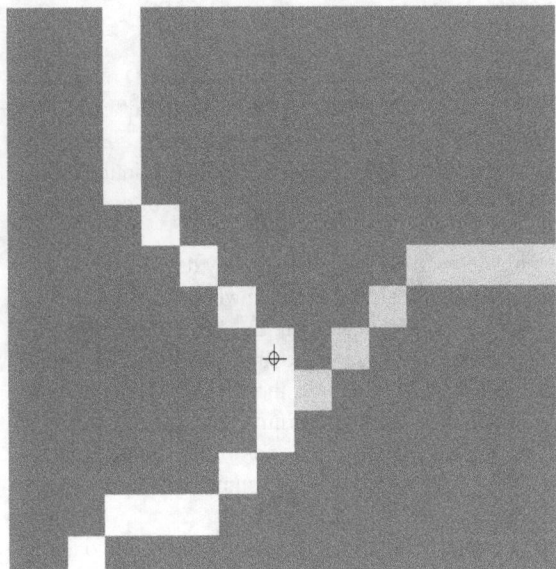

Figure 67. The field of view surrounding the pour-point point that should be used for delineating the AOI.

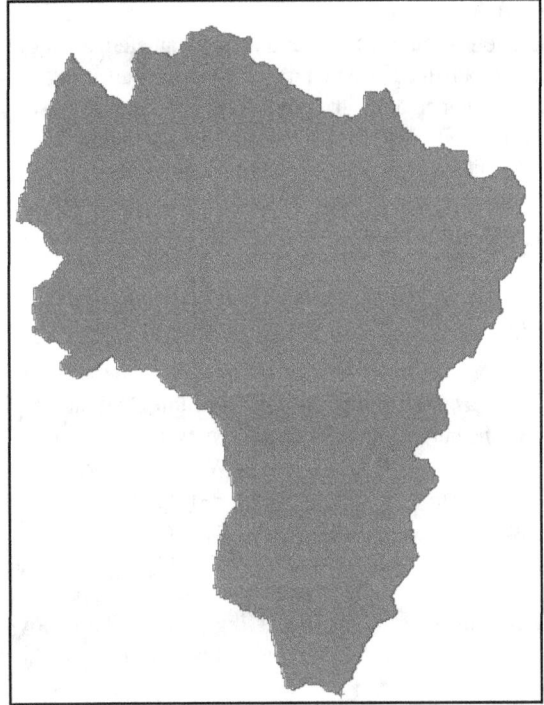

Figure 68. The AOI.

gage location (and, therefore, the location of the outlet of the contributing area). Press **Pan/Zoom>>Vector Control** (fig. 6) and specify D:\Workspace\weasel as the **Source Directory**. In the **Vector Control>>Cover** list, select demo_gages. **Vector Control>>Symbol** list, select the red cross by clicking on it with the left mouse button. Press the **Vector Control>>Apply** button. The **Map Display** should show a set of red cross symbols superimposed on the display of the DEM. The point at the bottom of the screen will be used as the gage of interest and the output or pour-point of the watershed.

Press the **AOI Delineation>>Digitize 'pour-point'**… button. The **Map Display** should be refreshed. Although the points from the demo_gages cover will still be visible, the DEM display will have been replaced by the Flow Accumulation surface. This surface will be almost entirely gray, except for the cells where the flow concentrates. The white cells will form what looks like a drainage network. Rather than trying to select the appropriate cell at this scale (which is nearly impossible to do successfully), zoom in first. Press the **Pan/Zoom>>Extent** button. Move the mouse cursor over the **Map Display** window and click two points with the left mouse button that define the opposite corners of a box around the confluence near the point at the bottom of the window. An area like the one shown in figure 67 should be on the screen. If the field of view is slightly different, it doesn't matter as long as the cell that is marked with cross-hairs in figure 67 can be accurately selected. To refresh the display with the fully enlarged field of view, press the **Pan/Zoom>>Full View** button.

Press the **Okay** button on the **Locate Pour-Point** menu. Now move the mouse cursor over the cell marked in figure 67 and click with the left mouse button to select the pour point. The mark will be a red cross and will be slightly lower. Be sure to avoid clicking on the cell below the confluence because this will result in an AOI that also includes the contributing area associated with the tributary. The **Map Display** window will be updated by zooming out to the full extent of the Analysis Window and showing the contributing area above the selected point. A map like the one shown in figure 68 should appear.

The area reported on the **Message Board** should equal approximately 289 square miles. If the image or the reported area does not match, then restart the **AOI Delineation>>Digitize 'pour-point'**… tool and try again. If the results do match what is being described here, then press the **AOI Delineation>>Apply** button to proceed. The result of this will be a new GRID, called aoi, that is stored in the Write Directory. In addition, there will be a coverage, called aoi_v, stored in the Write Directory. Do not move, rename, or destroy either of these geodatasets.

After this, GRIDs of slope and aspect will be automatically created and stored at D:\Workspace\wk_demo\output\surfaces\dem\filled\slope\grid and D:\Workspace\wk_demo\output\surfaces\dem\filled\aspect\grid, respectively. This is the final step in the setup phase.

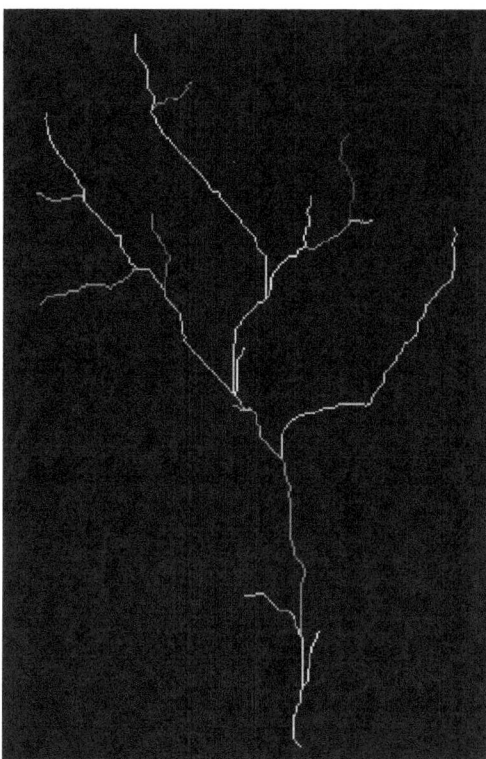

Figure 69.   The normal-type drainage network, based on a minimum threshold of 2000 upslope cells.

The Delineation Phase

During the parameterization phase, the GRID, `aoi`, will be associated with the geographic feature type (through the **Tool Panel>>Parameterization>> Parameterization Control Panel>>Geog Feat Type** too) called `one` and will be used derive parameters that describe the entire watershed to the PRMS model. Now zone maps need to be derived for the remaining types of geographic features. As specified in the description of the modeling application, there are four more types: `nhru`, `nradpl`, `nssr`, and `ngwres`. The first to be discussed will be the derivation of the `nhru` (that is, HRU) zone map.

As mentioned above, HRUs are normally conceived of as the contributing area associated with one of the banks of a link in the drainage network. Therefore, prior to delineating HRUs, a map of the links in the drainage network needs to be created. First, make sure that the **Tool Panel>>Current Zone Map** (fig. 23) specified is `aoi`. Then press the **Tool Panel>>Delineation of features>>Automated Methods** button. Next press the **Tool Panel>>…>>Drainage networks>>(normal)** button (fig. 29) to start delineating the drainage network. Type `2000` into the **Drainage Extraction>>Threshold>>(Cell Count)** slot (fig. 30) and press `<enter>`. Then press the **Drainage Extraction>>Apply** button. On the **Output Zone Map** menu (fig. 28), name the output `str2k`. Leave the **Mask…** and **Pack…** checkboxes in the default positions. Having the **Mask…** checkbox in the *On* position will eliminate any drainage links that are not overlain by cells in the Current Zone Map, `aoi`. Press the **Output Zone Map>>Apply** button. `str2k` should now be shown in the main **Map Display** area, and look like what is shown in figure 69.

The number of links in `str2k` can be checked by looking at the VAT. To do this, press **Pan/Zoom>>Raster Control** (fig. 6) and select `str2k` from the **Raster Control>>Input Grid** list (fig. 12). If it is not present, verify that the Source Directory is pointing to `D:\Workspace\wk_demo`. Once the `str2k` name has been clicked on in the **Input Grid** list, the **Raster Control>>Feature Tables** list should be updated. Highlight `STR2K.VAT` and press the **Raster Control>>List Table** button. The resultant table should show that there are 27 links. Notice that there are several zones that have a COUNT of only 7. One zone only has a COUNT of 1. This COUNT value shows the number of cells within the zone. While such small zones (that is, links in the drainage network) are not a problem for the PRMS application being described here, users with models that explicitly route water on the basis of stream length should be aware of extremely short links in the drainage network. Corrective action may be necessary, depending on the model being run. Press the **Quit** button.

Now that the drainage network has been constructed, the contributing areas associated with each bank (referred to as *left* and *right* banks here) of the links can be derived. Make sure that the **Tool Panel>>Current Zone Map** (fig. 23) specifies `str2k` before pressing the **Tool Panel>>Automated Methods** button. Then press the **Delineation of Features…>>Contributing areas>>two** button (fig. 29). The GIS Weasel will work on this for a small amount of time and then ask the user to name the output. On the **Output Zone Map** menu (fig. 28), name the output `str2ktp`. Leave the **Mask…** and **Pack…** checkboxes in the default positions. Setting the **Mask…** checkbox to the *Off* position will maintain contributing areas even if the areas are not overlain by cells in the Current Zone Map, `str2k` (which is desired here). If the output zone map matches the silhouette of the drainage network (that is, `str2k`), then the **Mask…** checkbox was likely in the *On* position when the output was being named. If so, redelineate the map. If not, press the **Output Zone Map>>Apply** button. `str2ktp` should now be shown in the main **Map Display** area and look like what is shown in figure 70. `str2ktp` is stored in the user's Write Directory. Another copy of this data set will be stored in the Output Subdirectories (described in the "Advanced Topics" section) at ...`output/zones/str2k/two-plane/grid`.

The number of zones in `str2ktp` can be checked by looking at the VAT associated with that GRID. First select the GRID via the **Pan/Zoom>>Raster Control>>Input Grid** list (fig. 12), then select the VAT from the **Feature Tables** list and press the **List Table** button. Sixty rows are in this table. Notice that some of the COUNT values are very low. Looking at the VAT is a suggested strategy for detecting such small units because small units are difficult to visually detect in the **Map Display** window.

Although extremely small HRU areas will not cause a computational problem for PRMS, most modelers feel that having such small units implies a false degree of precision about the input data and the interpretations that can be made on the basis of these data. The uncertainty associated with such small units pertains to the shape of such a small HRU and, more importantly, the values of the parameters associated with the HRU. This uncertainty propagates forward to the results of the environmental model and should be recognized and handled appropriately by the modeler.

A minimum HRU area of half a square kilometer is suggested as a rule of thumb. This threshold area may increase as the area of the AOI increases. Given a DEM with a cell size of approximately 100 meters, this equates to about 50 cells. The removal of spuriously small zones is carried out by using the **Tool Panel>>Dissolve** tool (fig. 23). Such zones are given the identifier of the neighboring zone (whose area exceeds the threshold) with the longest shared border. In the case of zones with very small cell counts, several neighbors can have equally long borders with a spuriously small zone. Therefore, the reassignment should be considered a random selection process. Again, examining the VAT can be helpful for selecting a minimum number of cells for a zone. If there are zones that the user wants to merge with a specific neighbor, then the **Tool Panel>>Id # manipulation** tool should be used. The user should now start the **Tool Panel>>Dissolve** tool and specify the **Dissolve Zones>>Minimum Area>>(Cell Count)** of 50, then press **Dissolve Zones>>Apply**. The output should be named str2ktp2. The VAT should reveal that there are 49 zones in this new version of the zone map.

For this sample application, this zone map will be used as the representation of the HRUs (nhru). In other cases, the geomorphology or the dominant hydrological processes associated with a given application, this zone map may not be appropriate. For example, the pair of zones (green and tan) at the right extreme of figure 70 might be considered too long. The red zone and white zone pairing at the northern extreme of the AOI may not look like the two zones need to be grouped together. If this is the case, the user has a variety of options. Tools like **Tool Panel>>Contrib Area** and **Tool Panel>>Digitize** can be used to modify the map. The user may even opt to edit the drainage network map (str2k) with these same tools prior to using **Tool Panel>>Automated Methods>>Contributing Areas>>two** to make the map of left and right bank contributing areas.

Now that the finalized version of the zone map representing HRUs exists, the zone map of radiation planes (nradpl) can be derived from it. The user should ensure that the **Tool Panel>>Current Zone Map** is set to str2ktp2 prior to starting the **Tool Panel>>Delineation of Features>>Automated Methods>>PRMS Radiation Planes** tool (fig. 29). The output should be named str2ktp2rp. The user should not modify the radiation plane zone map. If the VAT is examined, it should be noted that the first zone has a cell COUNT of 1. This is intentional and should not be modified.

Because the HRU zone map (str2ktp2) also will be used to represent the subsurface reservoirs (nssr) and the groundwater reservoirs (ngwres), no more delineation is needed. The reader should note that the str2ktp2 zone map is not representing a simultaneous mix of these geographic feature types, with some zones representing HRUs, some representing sub-surface reservoirs, and some representing ground-water reservoirs. With respect to HRUs, every zone in str2ktp2 represents a different HRU. With respect to subsurface reservoirs, every zone in str2ktp2 represents a different subsurface reservoir. With respect to groundwater reservoirs, every zone in str2ktp2 represents a different groundwater reservoir.

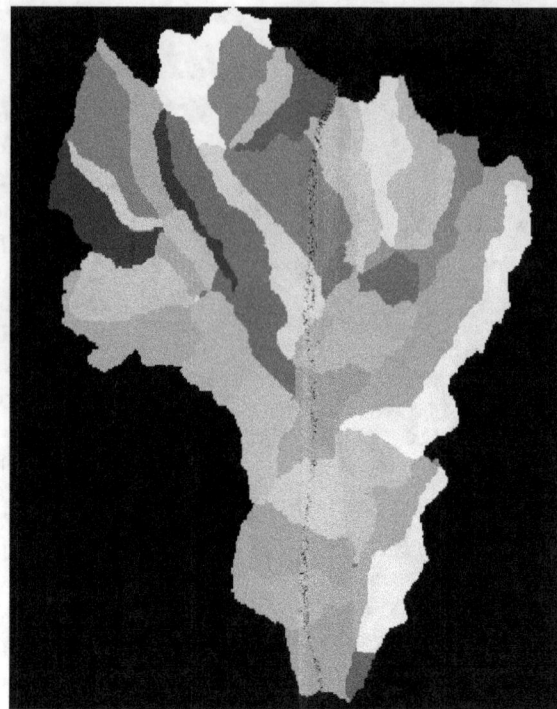

Figure 70. The left- and right-bank ("two-plane") contributing areas for the links in str2k.

The Parameterization Phase

Now that all the needed zone maps have been created, the user can start the parameterization phase by pressing the **Tool Panel>>Parameterization of Features** button (fig. 23). The first step that must be carried out is the assembly of a parameter settings list. Because PRMS is well known to the GIS Weasel development team, parameter settings specific to PRMS have been pre-packaged. To access these settings, the user should click on the *PRMS-topographic* option of the **Prepared Models>>Model** menu (fig. 63). Then click on the *PRMS-nontopographic_data_bin* option of the same menu. Answer *No* to the question of whether to abandon the current list of parameter settings.

Because many of these parameters rely on ancillary GRIDs found in the data bin (introduced in the "The Parameterization Phase" section, and fully explained in the "Advanced Topics" section), the user should now specify a value in the

Parameterization Control Panel>>Data Bin slot (fig. 62). The GIS Weasel is distributed with a sample data bin that corresponds to the same geographic area as that depicted in the sample DEM. The sample data bin is found within the GIS Weasel main directory (`weasel`). If the software has been installed in `D:\Workspace\weasel`, then specify `D:\Workspace\weasel\data_bin` in the **Parameterization Control Panel>>Data Bin** slot. Be sure to press the `<enter>` key after typing in the value. This will cause the GIS Weasel to check the validity of the data bin, which is required prior to parameterization. The "Advanced Topics" section provides details on how to respond to follow up questions and how to create an entirely new data bin for new geographic areas.

Next, the user needs to inform the GIS Weasel which zone maps are used to represent which geographic feature types. The current section has associated the English language terms like "sub-surface reservoir" and "radiation plane" with the names of specific zone maps (for example, `str2ktp2` and `str2ktp2rp`), but the GIS Weasel has no actual knowledge of these associations. All the GIS Weasel is aware of is a collection of zone maps, which are GRIDs of cells that have been arbitrarily assigned a name that may or may not match the names of the geographic feature types defined in the model. Looking back to the **Parameter Settings** menus, the user will see that parameters will be derived for three different geographic feature types: `one`, `nhru`, and `nradpl`. The user can make zone map specifications for these three types from the **Parameterization Control Panel>>Geog Feat Types>>Geographic Feature Type** list. The geographic feature `one` should be associated with the zone map called `aoi`. The geographic feature `nhru` should be associated with the zone map called `str2ktp2`. The geographic feature `nradpl` should be associated with the zone map called `str2ktp2rp`. When these associations have been made, press the **Feature Type/Method List>>Apply** button.

Although this section also introduced two geographic feature types, subsurface reservoirs and groundwater reservoirs, these types will not be used in the parameterization phase. These types were introduced to illustrate several ideas. The most important of which is the possibility of using the same zone map to represent multiple types of geographic features. The second is that although these types of geographic features are clearly the type of information that *could* be delineated and parameterized within a GIS, this is not always done. In this case, there are no ancillary GIS data available to generate parameter values for these geographic feature types. For instance, PRMS requires parameters that describe the linear and nonlinear rates at which each subsurface reservoir drains to the stream network. No such information is available in this sample application. Therefore, despite the possibility of delineating a zone map of subsurface reservoirs, this is not done because there is no way to generate the parameters.

When there are no GIS data for spatial parameters (or nonspatial ones), the user is required to gather values for these parameters by alternate means. This can include a literature search or discussions with topical or place-specific experts. This information needs to be provided to the model through the model parameter file. The version of PRMS being discussed here has been implemented within the MMS (Leavesley and others, 1996). The format of the parameter file is given by the MMS User's Manual. MMS provides parameter file editing capabilities that can be used to record manually determined parameter values for consumption by the model.

Theoretically, the **Parameterization Control Panel>>Apply** button could be pushed and the GIS Weasel would produce three standard ASCII files, one for the parameters associated with each geographic feature type. But for the purpose of supporting the PRMS model, as implemented within MMS, these standard files (referred to as *.par files in "The Parameterization Phase" section) are not sufficient. The standard files need to be converted into an MMS-format parameter file. The GIS Weasel will do this if the **Parameterization Control Panel>>MMS Parameter File** checkbox (fig. 62) is in the *On* position. The user should now confirm this.

When the GIS Weasel Parameterization Engine creates an MMS-format parameter file from its *.par files, it starts with a pre-existing MMS-format parameter file for the model in question. This is done so that the MMS-format parameter file produced by the GIS Weasel has all the information needed to allow the model to run, even if some of this information is not available within the GIS. Details on how this conversion is carried out are provided in "The Parameterization Phase" section. The specification of the pre-existing MMS-format parameter file is made in the **Parameterization Control Panel>>MMS Input** slot (fig. 62). The user should right-click in this slot and navigate to `...\weasel\src_c\mms_param_files` and select the `prms-daily.mms` file, and then press the **Input MMS Parameter File>>Apply** button.

The user should also specify the name of the MMS parameter file that will be the product by typing in the name of a nonexistent file in the **Parameterization Control Panel>>MMS Output** slot. If no pathname is specified as part of this value, then the file will be created within the Write Directory. If a pre-existing filename is specified, then the original will be destroyed. The user should not reuse the same name for the **MMS Input** and **MMS Output** values.

Once all configuration of the Parameterization Engine has been carried out, the user should press the **Parameterization Control Panel>>Apply** button to begin the generation of parameter information. A menu showing how far the parameterization process has progressed (in percent) will appear. When the parameterization process is complete, three new `Notepad` windows (or xterms running `vi`, for non-Windows users) will be presented. Each window will display one of the *.par files (for `one`, `nhru`, and `radpl` parameters). At this time, the *.par files will be used to generate the **MMS Output** parameter file.

The resultant MMS parameter file is ready to load into the model for GIS Weasel users operating on a Unix computing platform. Because of the way ArcInfo interacts with the Windows operating system, this file is DOS formatted (which affects how line feeds are encoded into the file). MMS models read only Unix formatted files. Therefore, Windows-based users need to convert this file from a DOS to a Unix format. This can be done any number of ways outside of the GIS Weasel. One example, for users that have CYGWIN installed on their computer, is to use `vi` to edit the file. Once inside of `vi`, typing:

```
:set fileformat=unix
:wq
```

will convert the format, save the changes, and exit the file editor. The resultant MMS parameter file can then be used within the model.

Advanced Topics

This section is intended for more advanced users to fully understand and exploit the GIS Weasel. The topics included here include: The Output Subdirectories, Reuse of Pre-existing Output, The **Pan/Zoom>>Clean WriteDir** Tool, The **Tool Panel>>Other Output** tools, The Data Bin, Plug-Ins, and Writing Parameterization Methods. The topics are not ordered according to importance or any other sequence. It is presumed that the reader has read and understood all the content of the previous sections in this manual.

The Output Subdirectories

This topic is intended to help programmers implement their own Arc Macro Language (AML) routines for use within the GIS Weasel. Any new routines should follow GIS Weasel conventions about the management of GRID outputs that is defined here. Understanding how the GIS Weasel manages output also may be of interest to a user who wants to inspect GRIDs produced for their AOI during a GIS Weasel processing session. The discussion of the different types of GRIDs (surfaces, zones) in the first section is built upon for the following topic. In addition, the reader should have a general understanding about directory structures.

Three types of routines are differentiated here on the basis of the type of GRID output made. These are surface-generating routines, zone-generating routines, and parameter-generating routines (that is, parameterization methods). The surface-generating routines are named with a `sfc_` prefix, the zone-generating routines are named with a `zone_` prefix, and the parameter-generating routines are named with a `param_` prefix. These routines are found in the `...weasel/routines/sfcs`, `...weasel/routines/zones`, and `...weasel/routines/params` directories, respectively.

The GRIDs produced by these routines are stored in a subdirectory structure, referred to as the *Output Subdirectories*. The Output Subdirectories can be thought of as a window into the processing sequences of the GIS Weasel routines. The GIS Weasel routines have built-in assumptions about the organization of the Output Subdirectories and can fail if that organization is not followed. In general, the user should avoid modifying the Output Subdirectories where possible. To remove any GRIDs from the Output Subdirectories, the **Pan/Zoom>>Clean WriteDir** tool should be used. This tool is explained later in this section. In addition to making programming of new routines easier, the Output Subdirectories form a metadata record of both the routines that were run, as well as, the inputs to and outputs from these routines.

Top-Level Areas within the Output Subdirectories

The top-level of the Output Subdirectories is named `output`. This subdirectory resides within the Write Directory. Depending on what has occurred within the GIS Weasel processing session, the `output` subdirectory can contain the following subdirectories: `surfaces`, `zones`, `files`, `combines`, `tables`, and `info`. The `surfaces` and `zones` areas are the most important to both the programmer and the user, and will be therefore the focus of most of this topic. The `files`, `combines`, `tables`, and `info` are generated by the Parameterization Engine and should not be modified. These files are explained briefly at the end of this section, under the sub-heading of *Output Subdirectories and the Parameterization Engine*.

The surfaces and zones Areas

Surface-generating routines take a pre-existing GRID surface as input and produce a GRID surface as output. This output is stored within the surfaces subdirectory, and the output named according to both the name of the input GRID and the routine. An example of this is the output from the routine sfc_aspect.aml. If the input to this routine is a GRID of elevation named my_dem, the output then will be stored in output/surfaces/my_dem/aspect/grid. There are several characteristics to note here. First, the subdirectory below surfaces, my_dem, is a container in which all GRIDs derived from my_dem are stored. This does not signify that the my_dem GRID actually exists at this location. In fact, the GRID could be in the Write Directory or anywhere else on the computer hard drives.

The name of the next subdirectory down in the path, aspect, refers to the name of the routine that was applied to my_dem. The sfc_ prefix of the routine name was dropped because the output is within the output/surfaces subdirectory, which indicates that the type of routine used was a surface generator (which, by convention, has a sfc_ prefix). In addition, the .aml suffix was not used, as all routines are AML scripts and the inclusion of this information in the Output Subdirectories yields no information.

At the very end of the pathname is the actual GRID that is produced by the application of the sfc_aspect.aml to my_dem. The name of this output is simply grid. The GRID output by any routine within the GIS Weasel is always named grid. The decision was made to embed metadata about the generation of a GRID into the naming of the subdirectories under which it is stored rather than in the name of the GRID. While GRIDs must have names of 14 letters or less, subdirectories have no such limit.

Zone-generating routines manage the output in the same way, except that the output is stored within the zones subdirectory. An example of this is the output from the routine, zone_one-plane.aml. For this example, assume that the routine is applied to a zone map named drain_network. The output would then be stored in output/zones/drain_network/one-plane/grid. The output from parameter-generating routines are managed in an identical manner to that of zone-generating routines (that is, it is stored in output/zones).

In effect, a metadata record is formed by the placement of a derivative of a surface or a zone map beneath a directory named after that surface or a zone map. In addition, the name of the routine that derived the information always can be ascertained from the path to a GRID stored in the Output Subdirectory. To restate, any GRID beneath output/surfaces/my_dem will have been created by the application of a surface-generating routine to a GRID called my_dem. The next subdirectory in the path, say aspect, can be used to resolve the actual name of the surface-generating routine. In this case, that would be sfc_aspect.aml.

In the case of output found beneath the zones area, there are two possible prefixes for routines, zone_ and param_. Determining whether a grid found in the the output/zones area was produced by a param_ or zone_ -prefixed routine is usually a process of elimination. In the few cases where there are both a param_ and a zone_ routine that reference the same output, the param_ routine simply runs the zone_ version. This redundancy is exclusively for the benefit of the Parameterization Engine (which only deals with parameterization methods). All the functionality of the zone_ routine could be moved to the param_ routine, and the zone_ routine eliminated. The choice to implement both a param_ and a zone_ routine is a style preference left to the programmer. Various menus in the GIS Weasel are designed to show only routines with specific prefixes, which might affect a programmer's decision.

A more complex example would be output/surfaces/my_dem/filled/flow-direction/flow-accumulation/grid. This pathname tells us that the grid was derived by the routine, sfc_flow-accumulation.aml by using the output from the routine sfc_flow-direction.aml. The sfc_flow-direction.aml, in turn, was applied to the output from sfc_filled.aml. sfc_filled.aml is dependent on a GRID called my_dem.

A benefit of the placement of routine output GRID below a directory named after the surface or zone map that the method was applied to is that it allows the storage of outputs from multiple applications of a single routine (to different inputs) with no confusion. For example, if sfc_aspect.aml also was applied to my_dem and your_dem, then this output would be stored in output/surfaces/my_dem/aspect/grid and output/surfaces/your_dem/aspect/grid, respectively.

Programmers should note that although the use of this naming convention may not be required for the routine to run successfully as a stand-alone program (for example, from the **Pan/Zoom>>Command Line** menu), all GIS Weasel routines and the Parameterization Engine find GRIDs by using this approach. If the data are to be accessible to other routines, then the data should be stored according to this convention. Also be aware that if this convention is not followed, GRIDs may be created in locations that other routines will try to write to or use, and problems may arise.

Output Subdirectories and the Parameterization Engine

Here each of the last four top-level areas of the Output Subdirectories are briefly introduced. The files subdirectory was introduced in the section on the parameterization phase. In general, files contains the ASCII parameter file output that can be used by the user's environmental model. Log files describing the derivation of the parameters also are stored here. The ASCII parameter files are always overwritten and are never reused by the Parameterization Engine. The `combines` and `tables` directories are created as part of the process of manufacturing the ASCII parameter output. When a GRID of parameters is derived (in the `zones` subdirectory), each cell records the derived parameter value. What the cells of such a GRID do not contain is the identifier of the zone that the parameter value describes. The `combines` subdirectory is used to reassociate parameter values with the identifier of the zone that the values describe. The `tables` subdirectory is used to aggregate all the parameters associated with a specific *geographic feature type* (this term is introduced in the "Introduction" section and is revisited in the "The Parameterization Phase" section). This is done by using the content found in the `combines` subdirectory. Programmers implementing their own delineation tools or parameterization methods should avoid interacting with these three areas because these areas are all created and managed by the Parameterization Engine.

The `info` area is created because ArcInfo uses specialized directories called *workspaces*. Part of a workspace is an `info` subdirectory. The `output` directory is a workspace and, therefore, has an `info` subdirectory. The `info` subdirectory is not actively used, but should not be modified by the user.

Reuse of Pre-existing Output

When a routine is run, one of the first things it does is determine where the output will be stored. The next thing it does is to determine whether this output already exists. If the output already exists, the routine will print a message to the screen to this effect and terminate. If it does not, then the routine will derive the output. This design promotes the automatic reuse of previously created GRID products. In these cases, the person or the software that ran the routine will not receive notification as to whether the output already existed. This feature is for processing efficiency.

In cases where a user does not want to reuse the pre-existing output when a routine is rerun, it is the responsibility of the user to make sure that the output from the routine to be rerun does not already exist. With the **Pan/Zoom>>Clean WriteDir** tool, a user can select one or more GRIDs to destroy prior to rerunning the routine. The mechanics of using this tool are described in the next section. The user can use the naming convention of the Output Subdirectories, defined above, to help select which pre-existing outputs to destroy.

A strong word of caution is warranted here. The elimination of the output from a routine does not guarantee that *no* pre-existing GRIDs will be used in the subsequent execution of that routine. A routine may run any number of other routines, which may, in turn, run other routines. For example, if a user ensures that the output of `param_method1.aml` has been destroyed, then the output of this routine will be derived anew the next time it is run. If `param_method1.aml` actually just multiplies the output from `param_method2.aml` by 10 and the output of `param_method2.aml` already exists, then rerunning `param_method1.aml` will simply be manipulating the pre-existing output of `param_method2.aml`. It may be the case that the user also wants to ensure that the output of `param_method2.aml` also has been destroyed.

If the user wants to avoid the reuse of any pre-existing outputs, they need to resolve all the subordinate routines (that is, by reading the AML scripts) and eradicate all of the outputs of these routines. Needless to say, this requires a high degree of expertise on the part of the user. A much less labor-intensive option, although extremely aggressive, is to use the **Pan/Zoom>>Clean WriteDir>>Nuke** tool, described in the next section.

The authors of the GIS Weasel have considered writing tools to resolve all the subroutines of a routine, but have decided against it because of the relatively weak support for this kind of programming within AML. An additional rationale for this decision is that even if the Output Subdirectories are almost empty, the entire parameterization process for an AOI is usually a quick process (on the order of 10 minutes or less).

As an additional aid to the user, the Parameterization Engine has been enhanced to point out if the user is about to reuse pre-existing output in the parameterization effort. It should be noted that this feature will only check for the existence of the outputs to the parameterization methods themselves. It will not look for the outputs of any subroutines that might be run by the parameterization methods. After the user presses the **Parameterization Control Panel>>Apply** button, the Parameterization Engine will check for pre-existing outputs. If any are found, an informational menu will be presented to the user. At this point, the user can continue with parameterization or stop and remove these pre-existing products.

The Pan/Zoom>>Clean WriteDir Tool

The **Pan/Zoom>>Clean WriteDir** tool was not discussed in detail in the section on the delineation phase, where it is first mentioned, because the authors wanted to ensure that the logic behind the Output Subdirectories had been fully discussed prior to introduction. The primary purpose of this tool is to allow the user to eliminate GRIDs that are no longer needed from the computer disk drive. The operation of the tool is very simple.

The **Product Delete List,** on the right of the **Clean WriteDir** menu shown in figure 71, specifies all files and GRIDs that will be destroyed if the user presses the **Apply** button. By default, this list is populated with the names of any GRIDs found in the Write Directory that have three consecutive zeros in their name. These GRIDs are assumed to be "scratch" GRIDs that were left behind by a routine that did not properly clean up or a routine that was interrupted prior to completion.

The output from a routine can be added to the **Product Delete List** by clicking on its name in the **All OUTPUT Product List** with the left mouse button. The **All OUTPUT Product List** shows the names of the Output Subdirectories created by all previously executed routines. The actual name "grid" is not shown (remember that all GIS Weasel-created outputs, although named "grid", are stored in different directory locations). Both the grid and the subdirectory that encloses it will be destroyed when the **Apply** button is pushed.

If a directory is added to the **Product Delete List** and destroyed, then any subdirectories to that directory also will be destroyed. For example, if output/zones/my_streams is destroyed, then output/zones/my_streams/ elevation-median also is destroyed. This can be a convenient way of destroying all derivatives of a zone map, for example. This being said, the user should make their selections carefully.

Alternatively, the user can click in the **All OUTPUT Product List** with the right mouse button and specify a GRID using a **Grid Manager**. This option is useful for specifying GRIDs that are located someplace other than the Output Subdirectories, such as the Write Directory.

The **Nuke** button is intended to strip the Output Subdirectories of all but the bare minimum of data. The only thing that will be left behind after this button is pushed is the path output/surfaces/<dem>/filled/flow-direction/ flow-accumulation/ and the GRIDs that may exist at some of the levels of this path. The variable, <dem>, refers to the name of the Input Elevation Grid and that the filled portion of the path will not be present if the user declined to fill the pits in the elevation data set during the setup phase. This is a relatively drastic action to take, but this is a way the user can ensure that all routines are actually deriving output. Only restarting the GIS Weasel and creating an entirely new Write Directory is more aggressive. Users have asked for this tool for cases where their previous parameterization effort relied on a Data Bin (discussed in the "Data Bin" section) that has since been completely updated, and users want to be sure that any new parameterization results reflect their newly updated Data Bin. The operation of the **Nuke** button will not be affected by the specifications contained in the **Product Delete List**.

Names can be removed from the **Product Delete List** by clicking on them. The **Close** button will dismiss the **Clean Write-Dir** tool without destroying any data.

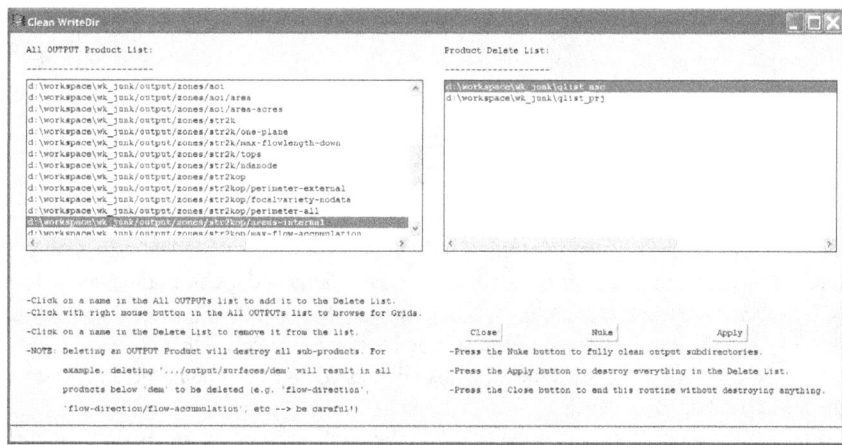

Figure 71. The **Clean WriteDir** tool.

The Tool Panel>>Other Output Tools

Most GIS Weasel routines have been associated with at least one component (button, menu, list) within the GIS Weasel GUI. The source code for these routines can be found in .../weasel/src_aml/routines. This area does not actually contain routines, but holds three directories that form functional groupings for (1) surface-generating routines, (2) zone-generating routines, and (3) parameter-generating routines. These are found in the sfcs, zones, and params subdirectories, respectively.

The GIS Weasel provides an alternate way for a user to access the surface- and zone-generating routines with the **Tool Panel>>Other Output** button (fig. 23). Surface-generating routines can be run through **Tool Panel>>Other Output>>Surface Generators**. Zone-generating routines can be run through **Tool Panel>>Other Output>>Zone Generators**. Parameter-generating routines are not accessible through this tool; those routines can be accessed with the **Tool Panel>>Parameterization>>Geographic Feature Type/Method List** tool.

Accessing routines in this way is essentially unsupported. Although the user is encouraged to explore the routines within the GIS Weasel package, the proper selection and execution of the relevant routines is the responsibility of the user. The **Tool Panel>>Other Output** tool is really just a way to ensure that the source code for all routines within the GIS Weasel package can be browsed by the user.

The Data Bin

For this topic, it is assumed that "The Data Bin" section in the "The Parameterization Phase" section and the "Introduction">>"Basic GIS Terminology" section have been read and understood. The authors assumes the reader has a grasp of database concepts and terminology. Further, the user needs some basic knowledge about the creation and manipulation of ArcInfo GRIDs. The current discussion is not intended to serve as an introduction for any of these subjects. It is intended to provide a sufficiently detailed description of the Data Bin and contents so that a user is able to replace any or all of parts of the Data Bin. This section will not document which tools or routines make use of the Data Bin. Regarding parameterization methods, the user is referred to the Appendix for information on which methods rely on the Data Bin.

The reader is reminded that the Data Bin is not necessarily required for a particular GIS Weasel processing session. In general, the Data Bin is used only during the parameterization phase. A user's application dictates which, if any, pieces of a Data Bin are required, usually by virtue of which parameterization methods (that is, the **Parameter Settings**) will be executed. The description presented here is of the maximum range of information that the GIS Weasel could possibly need.

There are three standard subdirectories within the Data Bin. These contain information relating the speciation and density of forest vegetation, a composite of land- cover information, and soils information. The three subdirectories are named forests, lcov_comp, and soils, respectively. Figure 72 provides a schematic view of these three areas, shown in red, and the contents.

The forests area contains two GRIDS, shown as blue boxes in figure 72. One is named density and contains values between 1 and 100. These values indicate vegetation density. This 1-km by 1-km cell-sized GRID data was derived by the U.S. Forest Service (Zhu, 1994) by using Advanced Very High Resolution Radiometry (AVHRR). The reader is referred to Zhu (1994) for more information on the derivation of this data. This GRID has no VAT. The values may be expressed as integer or floating-point numbers, although processing will assume integer values exist within the GRID. The second GRID, also published by the U.S. Forest Service (Powell and others, 1993; Zhu and Evans, 1994) is called lower48. The codes assigned to each cell in lower48 reflect the dominant type of forest species within that cell. The numerical codes and corresponding English equivalents are listed in figure 72 beneath the blue box labeled lower48. The numerical codes are stored within the lower48 VAT in the VALUE item.

Unfortunately, neither of these data sets are currently (2006) being distributed by the original authors. While the GIS Weasel authors have copies of these data, it is suggested that the user find alternate sources to replace these data. Newer, better data have been produced since (for example the 1992 National Land Cover Data Set (Vogelmann and others, 2001)). The only requirement for new data used in place of that described in figure 72 is that it mimics the formatting of the original GRIDs. This includes the names of the GRIDs and the directories (ArcInfo workspaces, to be precise) the GRIDs are stored in, the range of values within the GRIDs, and, in the case of categorical GRIDs, like lower48, the numerical codes refer to the same species listed in figure 72.

There are no assumptions within the GIS Weasel as to the coordinate system (projection, datum, and so on) of any of the GRIDs within the Data Bin, except that it is consistent with that of the Input Elevation Grid. There is no requirement on the numbers of rows or columns of data within the replacement Data Bin GRIDs, or the size of the GRID cells. The GRIDs being described here all have 1-km by 1-km cells, but this was dictated by the quality of the data available at the time the Data Bin was designed and implemented. The geographic region described by the GRIDs in the Data Bin should, of course, fully extend through the AOI.

The GRID of land cover, `lulc`, is a composite of the `lower48` data with version 2.0 of the North America Land Cover Characteristics Data Base (NALCC) (*http://edcsns17.cr.usgs.gov/glcc/tablambert_na.html*). Specifically, the compositing process maintained all original values of cells within `lower48`, except where code value `23` (that is, nonforest) was found. For these cells, the values from the corresponding NALCC cells were used. As a result, it can be seen from the code listing associated with `lulc` in figure 72 that the codes within `lulc` range from 1-23 and from 101-127. The original range of codes within the NALCC data was 1-27, but was augmented by 100 to avoid erroneous mixing with the code scheme of the `lower48` values. The codes for the `lulc` GRID are stored in a VAT by using the VALUE item.

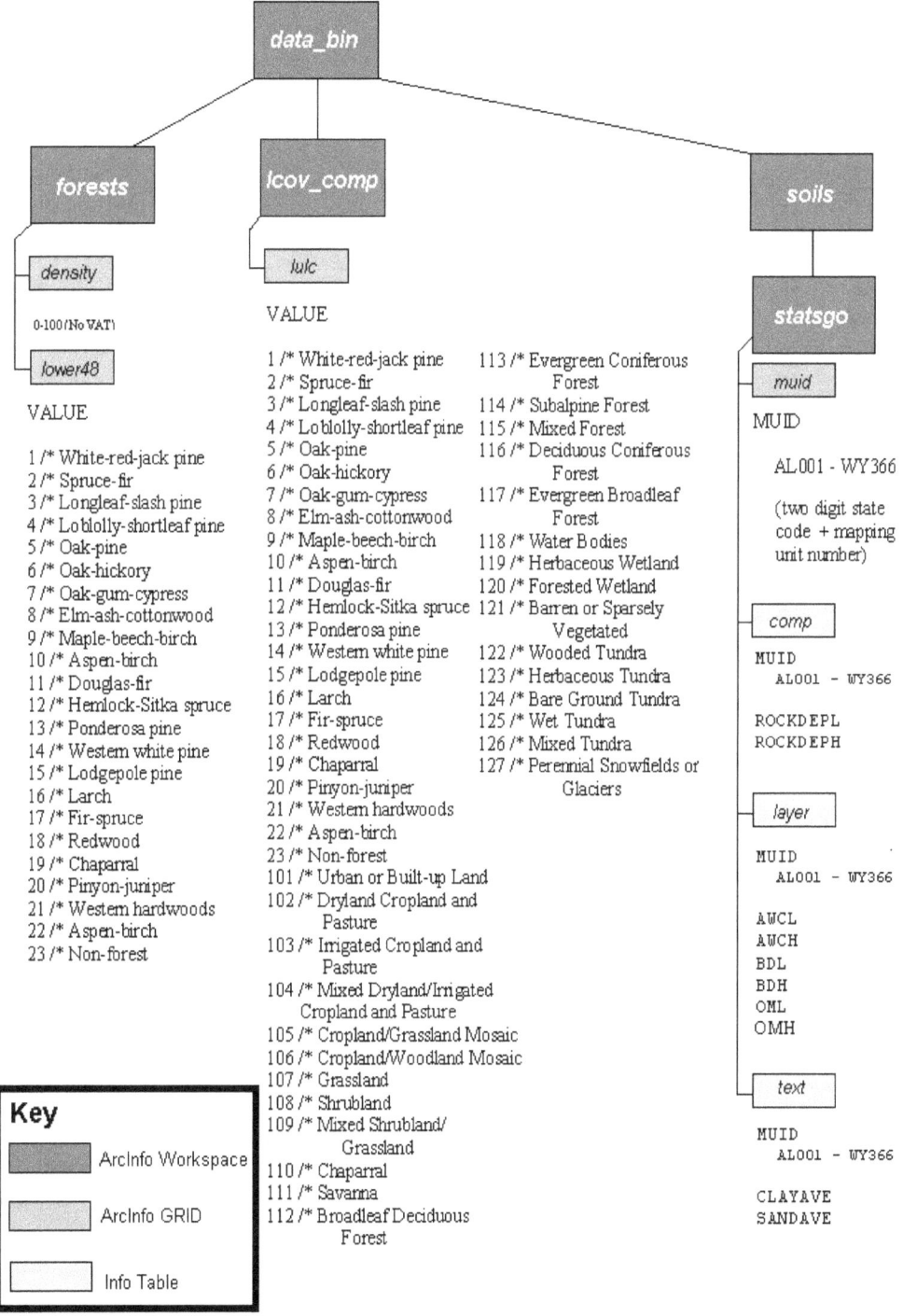

Figure 72.   A description of the Data Bin.

In the case of the creation of new `lower48` or `lulc` GRIDs, there is no requirement that instances of all codes should be present within the replacements. For instance, if code value `16` ("Larch") simply doesn't exist within the AOI that a Data Bin describes, then it will not cause problems for any of the GIS Weasel tools or routines. The requirement is that GRIDs introduce no new codes.

The final GRID in the Data Bin is a subset of a U.S. Geological Survey (USGS) version (Wolock, 1997, *http://water.usgs.gov/GIS/metadata/usgswrd/XML/muid.xml*) of the (STATSGO) soils data set. The STATSGO data set was originally published by the U.S. Department of Agriculture (1994, *http://www.ncgc.nrcs.usda.gov/products/datasets/statsgo/*). The most notable change made by the USGS version is that the data were encoded into a GRID format (instead of a vector format). Both versions of this data set are described as being suitable for application at "regional" scales (even more so than the 1-km cell-sized data already described.). The reader should avoid using these data to generate zone maps or parameters about areas whose scale is too fine relative to the STATSGO data.

The arrangement of the STATSGO GRID and related tables is relatively complex, although the USGS version has been simplified somewhat. Figure 72 shows that there is one GRID, called `muid` and three tables.

Within the `muid` GRID, spatially contiguous groupings of cells are uniquely identified as *mapping units* or *MUs*. This identification is made by using an alphanumeric field (column), called `MUID`, within the GRID VAT. It is important that any replacement of this GRID contain the `MUID` field in the VAT, because this field is used to relate to the three tables.

The three tables must be named `comp`, `layer`, and `text`. These tables must exist within the same `soils/statsgo` directory as the `muid` GRID for the GIS Weasel routines to be able to find the tables. There are many more tables available in both STATSGO versions. The presence of these extra tables within the Data Bin is acceptable, but not necessary for the operation of any GIS Weasel routines. Further, these tables, as defined by the USDA and USGS authors, have a much larger variety of fields than what is needed by the GIS Weasel. Again, the presence of this extra information is acceptable, but is not necessary.

The original USDA version of the `comp` table provides a way to describe more than one type of soil (called a *component*) within a mapping unit. A component is specified as a proportion of a mapping unit, but the location of the component is not specified. The motivation behind this is to represent the heterogeneity that is so inherent in soils in an aspatial way.

The USGS version of the STATSGO data maintains the `comp` table, but effectively merges all the components within each mapping unit. This is done by providing an area-weighted average for each mapping unit. The only information used from this table is the depth of the soil, as measured from the land surface to bedrock. This depth is expressed in inches. Two values are maintained for each mapping unit: the shallowest and deepest soil depth values, stored in fields labeled `ROCKDEPL` and `ROCKDEPH`, respectively.

A similar simplification is made within the USGS version of the `layer` table. The original USDA version of the `layer` table provides a way to discriminate amongst the horizons or strata of soil (*layers*) within the soil profile associated with a mapping unit. The USGS version provides a vertically averaged version of the characteristics associated with each layer, yielding a single value for each characteristic that describes the entire soil profile for each mapping unit. The average low and high values for available water holding capacity (`AWCL`, `AWCH`), bulk density (`BDL`, `BDH`), and organic matter (`OML`, `OMH`) are the fields used from this table by the GIS Weasel. `AWCL` and `AWCH` are expressed as a rate in inches/inch. The units of the `BDL` and `BDH` values are a percentage, as are those of the `OML` and `OMH` values.

The final table, `text`, contains soil texture information. The two fields within it that are used by the GIS Weasel are `CLAYAVE` and `SANDAVE`, which are average percentages of clay and sand content, respectively.

The tables are in an INFO format (see the **ArcDoc** for more information on INFO tables). All of these tables have, like the VAT associated with the `muid` GRID, an alphanumeric field called `muid`. This item is used by the GIS Weasel to relate (that is, logically connect) the nonspatial data content of the tables to the locations of the cells within the `muid` GRID, and, thereby, make this information accessible for spatial analyses. Strictly speaking, these codes are given by the original USDA STATSGO data product. For the purposes of the GIS Weasel, these codes can be any value, as long as those in the tables are logically consistent with those in the VAT of the `MUID` GRID.

The user is free to replace any of `muid` data sets in the Data Bin. As mentioned above, many newer, better resolution and quality geographic data sets have become available since the Data Bin was designed for use with the GIS Weasel. On the basis of the description in the preceding section, it is relatively easy to reformat newer data sets for the purpose of processing by the GIS Weasel. The most practical way to proceed with this exercise is to examine the specific details of the default Data Bin that is distributed with the GIS Weasel distribution package (found at `.../weasel/data_bin`). Although these data are not likely to have the same geographic extent of a user's AOI, the data will provide precise details on the formatting of the Data Bin contents.

Plug-Ins

Plug-ins are a way for a user to integrate customizations specific to user applications into the GIS Weasel. There is no requirement that any plug-in information has to actually be new software or data that are used by the GIS Weasel. Any type of information, such as text documents that describe recommended methodologies for the delineation of zone maps or Parameter Settings, can be included. In some cases, a large number of new tools or routines can be included in a plug-in. Third parties are free to publish (and document) plug-ins for the GIS Weasel. The GIS Weasel development team welcomes collaboration with such developers, but will not validate, endorse, document, or maintain these plug-ins. The GIS Weasel web site (*http://wwwbrr.cr.usgs.gov/weasel*) points to examples of externally created plug-ins This section will not document any specific plug-in. It will provide a description of how plug-ins can be used and created. How a plug-in is to be employed by the user shoud be documented by the plug-in developer..

Within the `weasel` directory that houses the GIS Weasel distribution package (the software, sample data sets, and start-up files), there is a subdirectory called `plugins`, where plug-ins are stored. Assume a user wants to add their own plug-in, called `my_plugin`. For this to be usable with the GIS Weasel, the user will need to create a directory called `weasel/plugins/my_plugin` and store all the customizing components in it. The `my_plugin` directory is the only thing that should be stored in the `weasel/plugins` area. For the remainder of this discussion, the term *plug-in* will be used to refer to a user's specific plug-in (such as `my_plugin`) and not the `weasel/plugins` area.

Plug-in Types

Software that is stored within the plug-in can be thought of as one of two types, referred to as *replacement* or *external* routines. Replacement routines are AML scripts with names matching those of routines within the standard GIS Weasel libraries (called *standard routines*). The intent of replacement routines is to have the GIS Weasel find and execute the plug-in version of a routine in lieu of the standard routine. Details on how to ensure that this replacement occurs are given in the next section. This can be a way of not only providing an alternative methodology to what is given by a standard routine, but also to have that alternative executed by or as part of pre-existing GIS Weasel tools. For example, if a user has a better way to derive the two-plane type of contributing area zone map and the user wants to ensure that it is run when they press on the **Tool Panel>>Automated Methods>>Contributing Areas>>Two** button, then the user/developer needs to name their implementation `two-plane.aml`, place it in their plug-in directory, and load that plug-in into the GIS Weasel.

External routines are any other kind of software. External routines can be written in AML or any other language and are not associated with any of the GIS Weasel GUI components. The user is, therefore, responsible for being aware of the existence of these routines and how and why the routines should be run, and making sure the plug-in is loaded into their GIS Weasel processing session. An external AML routine can be invoked directly by the user within a GIS Weasel processing session through **Pan/Zoom>>Command Line**. External AML routines run in this way are free to exploit a large variety of data, settings (for example, cell size, Analysis Window), and global variables associated with the GIS Weasel processing session, in the same way that replacement routines are. A set of external software can even have a GUI. It is typical to include documentation of external routines within the plug-in.

Using and Building a Plug-In

The GIS Weasel finds and runs core routines and tools by relying on the concept of a *path*. The path concept, used in most computing environments, including Unix and Windows, is essentially a list of directories. Whenever a user or a program specifies the name of a program to run, the computing environment will look into each directory in this list for a program whose name matches the name invoked. If the same program name exists in multiple directories of the path, the version whose enclosing directory is earlier in the list (that is, further left) will always be used. When a user "loads" a plug-in, the plug-in directory is actually attached to the path that the GIS Weasel uses, enabling the replacement of any pieces of the standard core of the GIS Weasel software without modifying the core software.

There are several methods by which a plug-in can be added to the path within a GIS Weasel processing session. The first is by pressing the **Write Directory and Input Elevation Grid Specification>>Plugins** button (fig. 4). As described in the section on the setup phase, this is the first menu that the user sees after starting the GIS Weasel. Loading a plug-in at this early stage of a GIS Weasel processing session is largely for the benefit of a plug-in that provides replacement software. The next place that plug-ins can be added to the GIS Weasel path is by typing `&run plugins.aml` into the **Pan/Zoom>>Command Line** menu (fig. 6). The **Pan/Zoom>>Command Line** menu is available through almost all parts of a GIS Weasel processing session. The third place that a plug-in may be added to the GIS Weasel processing environment is through the **Tool Panel>>Parameterization of Features>>Prepared Models>>>Plugins** button (fig. 63).

Each time any of these methods for loading plug-ins is used, the Plug-Ins menu shown in figure 73 is shown. Instructions for the use of this menu are presented in its top half. The **Plug-Ins available** list shows the names of all the subdirectories found in the .../weasel/plugins directory. A subdirectory can be pre-pended to the GIS Weasel path by clicking on its name in the **Plug-Ins available** list. When this is done, the selected plug-in name will appear in the **Plugins loaded** list. More than one

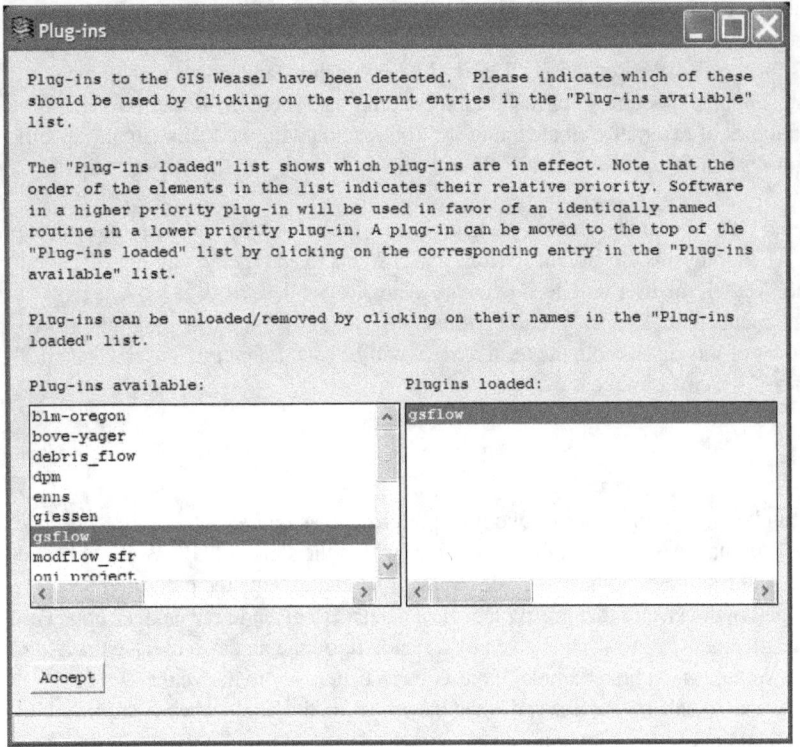

Figure 73. An example of the **Plug-ins** menu.

plug-in name can be selected. The order of the plug-in names in the **Plugins loaded** list defines the priority of the plug-ins. Plug-ins may be removed from the path by clicking on their names in the **Plugins loaded** list. As stated earlier, the purpose and usage of a a plug-in should be documented by its developer.

All three methods for adding plug-ins operate in an identical fashion, except that the **Tool Panel>>Parameter ization>>Prepared Models>>>Plugins** button has an additional feature. If a file named list.<plugin> (where <plugin> is the name of the plug-in) exists within the plug-in directory, then the Parameterization Engine will attempt to read this file into the **Parameter Settings** menu. Figure 74 shows an example of this file, referred to as the *parameter settings file*, for a plug-in called topnet. Even if there is no software in a plug-in, the parameter settings file can be a useful way to configure the list of standard GIS Weasel parameterization methods to apply and what labels to apply to the output.

There are several formatting requirements for the parameter settings file. Any text to the right of the characters / * is understood to be a comment and is ignored. The second line of the file in the figure 74 is the first actual parameter setting. This line consists of three pieces of information, separated by white space. The first is the name to attach to the output (that is, the parameter name). The second is the name of the geographic feature type. The third is the name of the parameterization method. No explicit pathname is associated with the parameterization method. This means that the path is used to find the routines. If there is a replacement param_area-meters.aml within the plug-in, then this routine will be found before the version that might exist in the core GIS Weasel parameterization library. Also note that the scroll bars on the bottom and right of the window may be needed to see all of the columns.

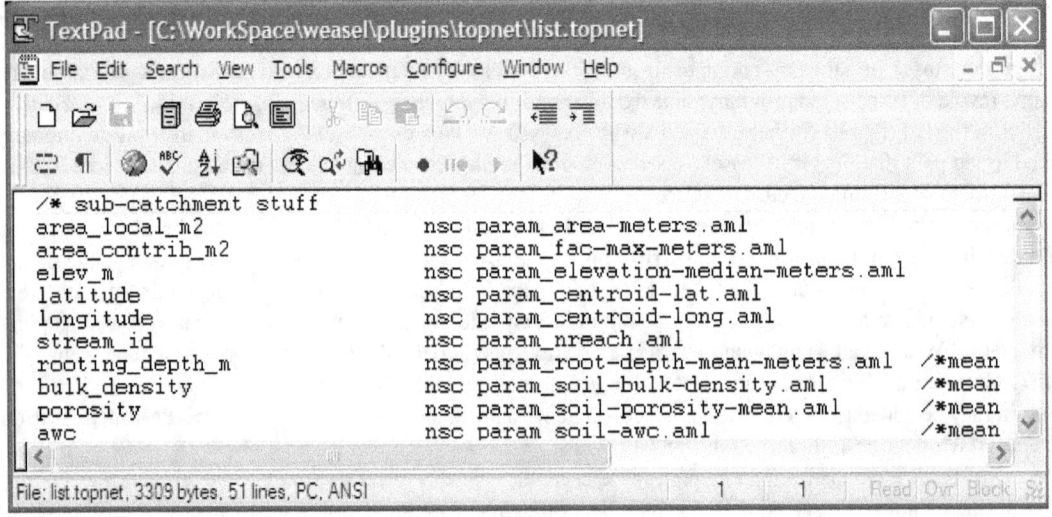

Figure 74. An example of a parameter settings file for the *topnet* plug-in.

Writing Parameterization Methods

This section is intended to serve as an introduction to the AML programming requirements and conventions associated with the GIS Weasel parameterization methods. It is not intended to serve as a reference on AML programming in general. The reader is assumed to be able to write AML scripts, to be familiar with raster data-processing concepts (such as Map Algebra), and to understand structured programming techniques. It is assumed that the section on the parameterization phase has been read. This section will present and discuss an example of AML code.

Introduction

Core GIS Weasel parameterization methods (that is, not part of a plug-in) are found in `weasel/src_aml/routines/params`. This area will be referred to here as the *parameterization library*. The reader is encouraged to open and examine several methods found here prior to reading the rest of this section. User-created parameterization methods should <u>not</u> be placed into the parameterization library; user-created parameterization methods should be placed into a plug-in (described in the previous section).

As mentioned above, a parameterization method must be coded to support a variety of required standards in order for the Parameterization Engine to be able to use it. Each parameterization method should produce a single, specially named GRID (that is, following the Output Subdirectories logic) as the final output. Only in rare instances does a parameterization method produce an ASCII output. As input, each method should accept at least two pieces of information: (1) the name of the zone map for which the parameter GRID is being derived, and (2) whether the method should be run in the "interactive" mode. The section on the parameterization phase described the interactive mode for parameterization methods when detailing the **Parameter Settings** menu (fig. 61). These input and output standards are minimum requirements for a parameterization method to function properly when used by the Parameterization Engine. Beyond this, how a parameterization method is actually implemented is completely open to the programmer. There are several conventions that the authors have found helpful, but these are optional.

Details on Standards

The relationship between the name of a parameterization method and where in the Output Subdirectories that method should place output (that is, the *pathname* of the output) was described earlier in this section. The programmer is reminded that if their method does not follow the Output Subdirectories logic, the Parameterization Engine will not be able to find the output GRID and use it to generate ASCII output.

Because of this logic, naming a parameterization method is somewhat important. The programmer should take care to avoid inadvertently choosing a name for their parameterization method that is already used in the standard parameterization library. Although such a naming conflict will not cause a problem for the Parameterization Engine, a user may find it confusing that more than one method places output into the same pathname. Additionally, the programmer should anticipate naming conflicts that might arise from the simultaneous use of multiple plug-ins. The programmer might elect to reuse the name of a parameterization method intentionally in order to replace a pre-existing routine (see the Plug-ins section for more on replacement software).

The Parameterization Engine assumes that the inputs to any method can be specified with an argument list. The argument list is defined on line 7 of the example method (fig. 75), with the `&args` directive. The first variable (also referred to as an *argument*) after `&args` in the list specifies the name of the zone map for which the parameter information is being derived. This first argument is usually named `dimname`. Because it only serves as a local variable, it can have whatever name the programmer desires. Whatever the local name for this first argument, it should always be used to reference the zone map within the method.

The second argument, usually named `inter`, is a Boolean variable indicating whether the method or any routines that the method might call should ask the user for input during execution. This was referred to in the section on the parameterization phase as the *interactive* mode. As with `dimname`, this variable can have any name. Regardless of whether a method has any options for user input, the Parameterization Engine will always provide a Boolean value for this argument. If the parameterization method calls other routines, especially those of the GIS Weasel core, the programmer is encouraged to pass the value of `inter` to those routines.

All subsequent references to line numbers and AML code examples will be in reference to figure 75.

Details on Conventions

This section will use the example in figure 75 to present a detailed description of a variety of programming conventions associated with parameterization methods. At the very top of this parameterization method are several lines of comments that include text such as the name of the method and author, the date of creation, and the purpose of the method. Immediately beneath that, on line 7 is the argument list that was described in the previous section. The argument list is a required feature of parameterization methods. On line 8 is a directive that indicates what should happen in the case that the method fails during execution. In this case, the &routine called return_error, found on lines 42-47, is executed prior to terminating the method. The most important aspect of this routine is the &return &error directive on line 46. Without it, the Parameterization Engine will receive no indication from a method if there was a problem with execution.

All core parameterization methods will check to see if the anticipated output already exists. If the output exists, then the method prints a brief message to the screen and closes execution. The code block shown in lines 12-15 implements this. Note the use of the convenience variable .param$path. The value is set by getpath_param.aml (on line 10) to output/zones/<dimname>, where <dimname> is the name of the zone map used as the first argument to the method. The variable, <dimname>, is the localized (that is, without a directory path) version of the %dimname% name. For example, if %dimname% (on line 17) was C:\Workspace\wk_demo\aoi, then %.param$path% would be output\zones\aoi.

On line 28, another routine, param_area.aml, is invoked. The argument list is %dimname% %inter%, as for all parameterization methods. On line 29, a local variable is set to point to the location where the GRID output by this method is expected. param_area.aml, is assumed to have followed the Output Subdirectories logic in placing the output. The local variable, area, is used in the remainder of the method as a convenience because it has fewer characters than %.param$path%/area/grid or output/zones/<dimname>/area/grid.

The programmer is encouraged to break their task into a series of simple methods, rather than having too many important GIS operations in a single large method. While the example in figure 75 is somewhat trivial, it illustrates that this can be done by having a method invoke and use the output of other routines. In addition, it demonstrates that creating "bite-sized" methods helps future implementation efforts because it promotes the subsequent reuse of routines like param_area.aml.

Rather than struggling to find a meaningful name for GRID that is about to be created in the Write Directory that does not conflict with names already found in the Write Directory, parameterization methods use a scratch name. This is encouraged for two reasons. Using fixed or hard-coded names for GRIDs that a method will create within the Write Directory is discouraged because this can easily lead to naming conflicts with data that a user might store in the Write Directory. Only the final GRID should be stored in the Output Subdirectories in order to keep this area as clean as possible. On line 31, a local variable is set by using the [scratchname] function. This function generates a name for a new geodataset that will not conflict with any already in the current Write Directory. The next line shows a Map Algebra expression where a new GRID with the name given by the %scratch% variable is created by multiplying the %area% GRID by a scalar conversion factor. The reader is referred to the **ArcDoc** or any GIS textbook that treats raster processing for more information on Map Algebra.

The getpath_param.aml is run (on line 34) for a second time. While not strictly necessary in this case, this is a good idea anytime the .param$path variable is about to be used. Subroutines can change this variable to values that may not be appropriate to the calling routine. In our example, .param$path was originally set on line 10 but a subroutine was run on line 28. It is safest to not make assumptions about whether or not the subroutine reset the .param$path variable.

On line 35, create_output_workspaces.aml is run. This routine ensures that the appropriate location for housing a final GRID output in the Output Subdirectories exists. On line 36, the %scratch% GRID is copied and renamed to grid in the process. After this copying operation, the original %scratch% GRID is removed (on line 37). Programmers are strongly encouraged to clean up after their routines and avoid cluttering the Write Directory with scratch GRIDs. If a scratch grid (called xx00000, for example) is used as input to a zone map or parameter-generating routine, the programmer should ensure that the method copies the result into the Write Directory and destroys the output area (output/zones/xx00000). Because the scratch name xx00000 is temporary, it is likely to be reused in subsequent processing. If this occurs, the GRIDs in the previously generated output area (i.e. output/zones/xx00000) are very likely to be erroneously reused.

The final important feature of the sample method is the &return statement on line 41, signaling the end of the method.

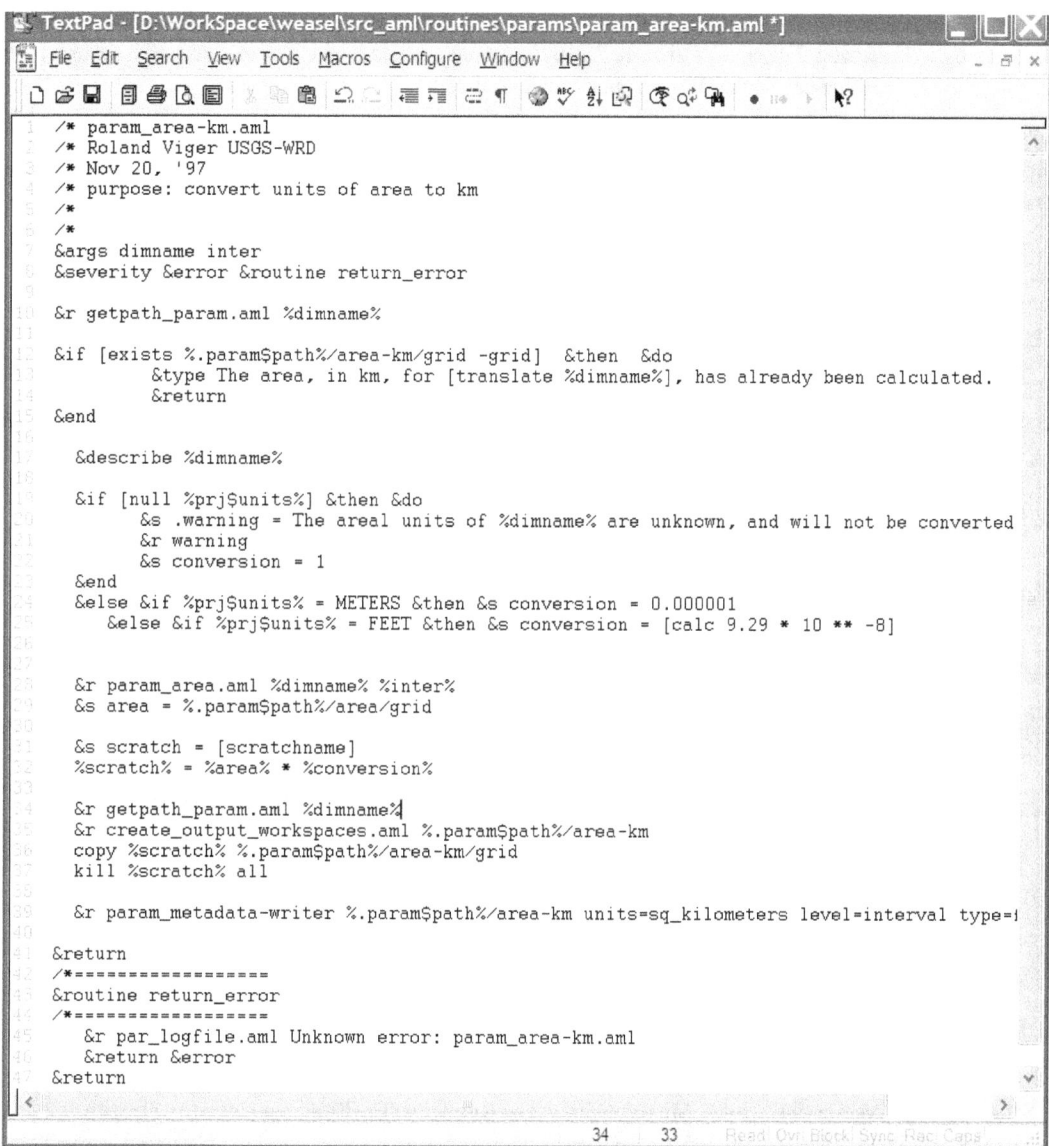

Figure 75. The `param_area-km.aml` parameterization method.

References

Anderson, J.R., Hardy, E.E., Roach, J.T., and Witmer, R.E., 1976, A land use and land cover classification system for use with remote sensor data: U.S. Geological Survey Professional Paper 964, 28 p.

Beven, K. J., Quinn, P.F., Romanowicz, R.J., Freer, James, Fisher, J.I., and Lamb, R., 1995, TOPMODEL and GRIDATB – A Users Guide to the Distribution Versions (95.02): Lancaster University, United Kingdom, Centre for Research on Environmental Systems and Statistics, Institute of Environmental and Biological Sciences, 31 p.

Elassal, A.A. and Caruso, V.M., 1983, Digital elevation models, USGS digital cartographic standards: U.S. Geological Survey Circular 895-B, 40 p.

ESRI, Inc., 2001, ArcInfo Workstation, Version 8.1 [software].

Hutchinson, M.F., Stein, J.A. and Stein, J.L., 2001, Upgrade of the 9 second Australian digital elevation model: Centre for Resource and Environmental Studies, Australian National University, accessed June 1, 2006 at *http://cres.anu.edu.au/dem/*.

Jenson, S.K. and Domingue, J.O., 1988. Extracting Topographic Structure from Digital Elevation Data for Geographic Information-System Analysis. Photogrammetric Engineering and Remote Sensing, v. 54, no. 11, p. 1593-1600.

Jobson, H.E., 1989, Users manual for an open-channel streamflow model based on the diffusion analogy: U.S. Geological Survey Water-Resources Investigations Report 89-4133, 73 p.

Leavesley, G.H., Lichty, R.W., Troutman, B.M., and Saindon, L.G., 1983, Precipitation-Runoff Modeling System - User's Manual: U.S. Geological Survey Water-Resources Investigations 83-4238, 207 p.

Leavesley, G.H., Restrepo, P.J., Markstrom, S.L., Dixon, M., and Stannard, L.G., 1996, The modular modeling system (MMS) User's manual (ver. 1.1): U.S. Geological Survey Open-File Report 96-151, 142 p.

Linsley, R.K., Kohler, M.A., Paulhus, J.L.H., 1982, Hydrology for Engineers (3rd ed.): New York, McGraw Hill, 508 p.

Miller, D.H., 1959, Transmission of insolation through pine forest canopy as it effects the melting of snow: Mitteilungen der Schweizerischen Anstalt für das forstliche Versuchwesen, Versuchsw. Mitt., v. 35, p. 35-79.

Powell, D.S., Faulkner, J.L., Darr, D.R., Zhu, Zhiliang, and MacCleery, D.W., 1993, Forest resources of the United States, 1992, General Technical Report RM-234 (revised): Fort Collins, Colorado, U.S. Department of Agriculture, Forest Service, 132 p.

Rawls, W.J., and Brakensiek, D.L., 1983 , Agricultural Management Effects on Soil Water Retention: Natural Resources Modeling Symposium, Pingree Park, Colorado, USA, October 16-21,1983 [Proceedings] p. 115-118.

Rosenberg, N.J. Blad, B.L., Verma, S.B., 1983, Microclimate - The Biological Environment (2nd ed.): Chicago, John Wiley & Sons, Inc., p. 170.

Snyder, J.P., 1987, Map projections - A working manual: U.S. Geological Survey Professional Paper 1395. 383 p.

Strahler, A. N., 1952, Dynamic basis of geomorphology: Geological Society of America Bulletin, vol. 63, no. 9, 923-938 p.

U.S. Department of Agriculture, 1994, State soil geographic (STATSGO) data base - data use information, (rev. ed.): Fort Worth, Texas, Natural Resources Conservation Service miscellaneous publication no. 1492, p. 212.

U.S. Department of Agriculture, Natural Resources Conservation Service, 2005, National Soul Survey Handbook, title 430-VI, [Online] accessed 10/27/2006. http://soils.usda.gov/technical/handbook/

U.S. Geological Survey, 1986, Standards for digital elevation models: U.S. Geological Survey Open-File Report 86-4, 46 p.

Vézina, P.E., and Péch, G.Y., 1964, Solar radiation beneath conifer canopies in relation to crown closure: Forest Science, v. 10, no. 4, p. 443-451.

Vogelmann, J.E., Howard, S.M., Yang, Limin, Larson, C.R., Wylie, B.K., Van Driel, Nick, 2001, Completion of the 1990s National Land Cover Dataset for the Coterminous United States from Landsat Thematic Mapper Data and Ancillary Data Sources: Photogrammetric Engineering and Remote Sensing, v. 67, no. 6, p. 650 - 662.

Wade, Tasha and Sommer, S.E. (eds.), 2006, A to Z GIS – An Illustrated Dictionary of Geographic Information Systems: Redlands, Ca., ESRI Press, 288 p.

Webb, R.M.T., Wolock, D.M., Linard, J.I., and Wieczorek, M.E., 2004, The Water, Energy, and Biogeochemical Model (WEB-MOD) – a TOPMODEL application developed within the Modular Modeling System [abs]: Eos. Transactions American Geophysical Union, Fall Meeting, in Supplements, Abstract H44B-03, v. 85 no. 47.

Wolock, D.M., 1997, STATSGO soil characteristics for the conterminous United States: U.S. Geological Survey Open-File Report 97-656, 17 p.

Zhu, Zhiliang, 1994, Forest density mapping in the lower 48 states – A regression procedure: New Orleans, La., U.S. Department of Agriculture, Forest Service, Southern Forest Experiment Station, Research Paper SO-280, 11 p.e

Zhu, Zhiliang, and Evans, D.L., 1994, U.S. forest type groups and predicted percent forest cover from AVHRR data: Photogrammetric Engineering and Remote Sensing, v. 60, no. 5, p. 525-531.

Glossary of Terms and Acronyms

A

AML   Arc Macro Language. The programming language of the ArcInfo Workstation GIS software. The GIS Weasel is chiefly written in this language.

AOI   Area of Interest. The geographic area that the data products from a GIS Weasel processing session will describe, depicted in an ArcInfo GRID.

ASCII   American Standard Code for Information Interchange. The de facto standard for the format of text files in computers and on the Internet that assigns a 7-bit binary number to each alphanumeric or special character. ASCII defines 128 possible characters (Wade and Sommer, 2006).

AWC   Available Water-holding Capacity. Available water capacity is the volume of water that should be available to plants if the soil, inclusive of fragments, were at field capacity. It is commonly estimated as the amount of water held between field capacity and wilting point, with corrections for salinity, fragments, and rooting depth (U.S. Department of Agriculture, Natural Resources Conservation Service, 2005).

B

BD   Bulk Density. Bulk density 1/10 bar or 1/3 bar is the oven-dried weight of the less than 2 millimeter (mm) soil material per unit volume of soil at a water tension of 1/10 bar or 1/3 bar. Bulk density influences plant growth and engineering applications. It is used to convert measurements from a weight basis to a volume basis. Within a family particle size class, bulk density is an indicator of how well plant roots are able to extend into the soil. Bulk density is used to calculate porosity. (U.S. Department of Agriculture, Natural Resources Conservation Service, 2005).

D

DEM   Digital Elevation Model. Originally referred as an ASCII data format for storing elevation data published by the U.S. Geological Survey (Elassal and Caruso, 1983). This term now is used to generally refer to any raster encoded elevation data set.

G

GIS   Geographic Information System. A system of software for the creating, storing, manipulation, visualization of, and analysis of digital geographically-referenced data. ArcInfo Workstation is an example of a GIS. ArcInfo Workstation is an example of a GIS.

H

HRU   Hydrologic Response Unit. This term refers to the primary spatial unit of the PRMS model (Leavesley and others, 1983).

O

OM   Organic Matter. Organic matter percent is the weight of decomposed plant and animal residue and expressed as a weight percentage of the soil material less than 2 mm in diameter. Organic matter influences the physical and chemical properties of soils far more than the proportion to the small quantities present would suggest. The organic fraction influences plant growth through its influence on soil properties. It encourages granulation and good tilth, increases porosity and lowers bulk density, promotes water infiltration, reduces plasticity and cohesion, and increases the available water capacity. (U.S. Department of Agriculture, Natural Resources Conservation Service, 2005).

P

PRMS   The Precipitation-Runoff Modeling System (Leavesley and others, 1983) is a watershed model.

T

TIN   Triangulated Irregular Network. A vector representation of a surface (usually elevation) using contiguous, nonoverlapping triangles, Triangles are formed based on irregularly spaced points, where points form triangle vertices (Wade and Sommer, 2006). This data model is not used by the GIS Weasel.

V

VAT   Value Attribute Table. This is a relational table that is sometimes associated with an ArcInfo GRID. Each row lists an identification number (labeled "value") assigned to one or more cells within the GRID, and the count of how many cells share that value.

Appendix: Parameterization Methods

This appendix is intended to serve as a reference describing each of the GIS Weasel parameterization methods. It is assumed that the reader is familiar with the full functionality and operation of the GIS Weasel, including the material discussed in the "Advanced Topics" section. Each description is intended to define the purpose of the method, but is not intended to provide details on implementation. The reader is encouraged to examine the source code for a method if they want more detail than what is provided here. All the parameterization methods are stored in `.../weasel/src_aml/routines/params`, where the "..." refers to the directory in which the GIS Weasel has been installed.

Introduction

A parameterization method is intended to derive some information about each zone in an input zone map. Parameterization methods do not usually make assumptions about what type of geographic features the input zone map depicts. As a result, the same parameterization method can usually be applied to more than one kind of input zone map, even within the same parameterization session. For instance, a parameterization method that determines the median elevation of each zone could be applied to a zone map of watersheds, a zone map of land-use patches, and a zone map of the links in a drainage network. If a parameterization method does make any assumptions about the content of the input zone map, it will be mentioned in the description in this appendix.

If the parameterization method requires other data to complete calculations, then these data are usually explicitly referred to within the method source code. To continue with the example from the preceding paragraph, the parameterization method would have a "hard-coded" reference to a GRID of elevation data within the source code. Such an explicit reference can be made using the actual pathname to a GRID, but is more frequently made using a variable that is set elsewhere in the GIS Weasel system. An additional example of a variable value that has been used in a large number of parameterization methods in this way is `.par$data_bin` that is set when the user specifies the location of the data bin in the **Parameterization Control Panel** (described in the "The Parameterization Phase" section). Parameterization methods can easily and robustly specify one of the standard GRIDs found in the data bin by providing the pathname to the GRID relative to the `.par$data_bin` variable (for example, `%.par$data_bin%/forests/density`).

If the user wants to run a parameterization method without the help of the Parameterization Engine, this can be done by typing from the **Pan/Zoom>>Command Line** interface:

```
&run param_<suffix>.aml <zone_map> <interactive>
```

where `<zone_map>` is the name of the ArcInfo GRID for which the results will be derived and `<interactive>` is set to either ."true." or ."false.." The `<interactive>` setting indicates if the user should be consulted if there is any possibility for specifying input to the parameterization method during the execution. `<suffix>` refers to the portion of the method name that will be used to name the location to which the output will be stored. All methods are invoked using these two arguments (including the `<interactive>` setting, even if there is no possibility for the user to specify execution-time inputs). The output of each method will be stored in Output Subdirectories (described in the "Advanced Topics" section) in an ArcInfo GRID named `.../output/zones/<zone_map>/<suffix>/grid`, where the "..." refers to the user's Write Directory. Running a parameterization method in this way will not yield a standard GIS Weasel *.par file (described in the "Parameterization Phase" section). If the user wants this GRID summarized in a standard GIS Weasel *.par file, then the parameterization method should be run through the **Tool Panel>>Parameterization** interface. This will yield both `.../output/zones/<zone_map>/<suffix>/grid` and the standard GIS Weasel `*.par` file.

Generic Parameterization Methods

Many of what are referred to as the "standard" GIS Weasel parameterization methods require additional ArcInfo GRIDs to complete processing. To support these methods, the data bin and contents (described in the "Advanced Topics" section) were designed and built from GIS data sets that were available for at least the lower 48 states of the United States. The standard methods are hard-coded to look for the data bin and contents. For a user whose AOI lies outside of the lower 48 states, the standard contents of the data bin will not overlap geographically. If the option of creating analogous data bin contents for the AOI is not feasible, then the standard parameterization methods that rely on the data bin are useless.

In order to provide some means of supporting users with AOIs located outside the conterminous United States, a large group of new parameterization methods were written. This group of methods is referred to here as "generic" parameterization methods. These methods are more simplistic than the standard counterparts. During execution, each generic parameterization method will prompt the user to specify the location of an ArcInfo GRID containing the information it needs. Generic parameterization methods merely take the mean or majority value of the cells in the user-supplied GRID (depending on the type of information that is expected) across each zone in the input zone map. The input GRID may contain values that vary on a per-cell basis, or may contain per-zone values. In either case, a summary of the information content in the input GRID is all that will be generated. No analysis is applied to derive new information. This effectively forces the user to create GRIDs content that directly corresponds to the definition of the parameter that the method is supposed to create.

These methods are all named using the convention `param_gen-<theme>.aml`, where `<theme>` refers to the information being derived. Despite the fact that all these routines have different names, the routines are computationally identical. The variations are limited only to the choice of summary statistic. Given the expectation that the level of measurement associated with the user-supplied GRID will be interval or ratio, a mean is taken. For information that is anticipated to be categorical, the modal (that is, majority) value is taken.

The real meaning of the output of a generic parameterization method is tied to the nature of the GRID that the user provides as input during the execution of a method. Although demanding that the user supply any such GRIDs places the significant burden, it at least allows the user to integrate their own data into a GIS Weasel parameterization session. The user is expected to examine the source code of the relevant generic parameterization methods prior to using the methods. Therefore, these routines will not be individually documented. The routine, param_generic.aml, is provided as a template to help illustrate how generic parameterization methods are constructed for those users seeking to implement their own generic routines (please also see the "Advanced Topics" section).

Format of Description

The layout and content of the description for each parameterization method has been written to follow a standard format as much as possible. The standard begins with the name of the parameterization method and an English-language *definition*, usually a single sentence. A more complete discussion is provided later in the *description* section of the entry. The definition is followed by a listing of the o*riginal purpose* of the method. Many of the GIS Weasel parameterization methods have been developed in support of specifically named and defined parameters for specific models. The original purpose specification attempts to capture this information. It is hoped that this information is useful not only to users of those particular models, but might provide useful cues to users of other, similar kinds of models in determining the relevance of a method for the user's needs. Next is a notation of *extra data used* to derive the output GRID. This list will not include the input zone map, but will include any extra GRIDs, coverages, INFO tables, and so forth, that might be used to derive the result. Following that is a list, labeled *subroutines called*, of any programs that are run by the parameterization method on behalf of the user during execution. The subroutine may, during execution, run other programs and that these other programs are not noted here. The user is encouraged to examine the source code of the subroutine to resolve the identity (if any) of any other programs that will be run in support of a given parameterization method. The description will discuss the basic details of the method and will, if necessary, expand on any special characteristics relating to it. The final portion of the description of a parameterization method is a list of *references*. This usually includes a reference to the simulation model that the method was developed to support. The user should note that the basic algorithms for statistics like mean, median, and standard deviation are provided by ArcInfo Workstation. No explicit reference is made to this product in the description of an individual method, but it is implied here for all routines.

Listing of Methods

param_area.aml

DEFINITION: Reports the area of each zone in the input zone map. Expressed in the XY units of the input GRID coordinate system.

ORIGINAL PURPOSE: (generic)

EXTRA DATA USED: (none)

SUBROUTINES CALLED: (none)

DESCRIPTION:

This method will determine the area for each zone in the input zone map. The areal units are not recorded in the output, other than in the description of the coordinate system of the output GRID.

REFERENCES: (none)

param_area-1st_order-smallest.aml

DEFINITION: Determines the identity of the smallest of the first-order drainage. The input zone map is expected to represent drainage areas.

ORIGINAL PURPOSE: (generic)

EXTRA DATA USED: (none)

SUBROUTINES CALLED:

```
sfc_flow-accumulation.aml
```

DESCRIPTION:

This method will first find the maximum flow accumulation associated with each zone in the input zone map. Then it determines the smallest of these maxima across the entire AOI. The zone whose maximum flow accumulation matches this AOI minimum is then extracted into the output GRID. The output GRID will only contain valid values for those cells whose location corresponds to those cells that constitute the smallest zone in the input zone map. The value for those cells in the output GRID will match those of the input.

If this method is run through the Parameterization Engine, then the smallest first-order zone will be the only one with a nonzero value assigned to it in the GIS Weasel *.par file. Also see `param_area-smallest.aml`.

REFERENCES: (none)

param_area-acres.aml

DEFINITION: Reports the area of each zone in the input zone map. Expressed in acres.

ORIGINAL PURPOSE: `hru_area` (PRMS), `basin_area` (PRMS, replaced by `param_area-total-nhru-acres.aml`)

EXTRA DATA USED: (none)

SUBROUTINES CALLED:

    `param_area.aml`

DESCRIPTION:

This method will determine the area by running the `param_area.aml` method and converting the units of the output to acres. If no units are specified in the input zone map coordinate system, then no unit conversion will occur and the results will be reported using the input zone map coordinate system.

REFERENCES:

Leavesley, G.H., Lichty, R.W., Troutman, B.M., and Saindon, L.G., 1983, Precipitation-Runoff Modeling System - User's Manual: U.S. Geological Survey Water-Resources Investigations 83-4238, 207 p.

param_area-hectares.aml

DEFINITION: Reports the area of each zone in the input zone map. Expressed in hectares.

ORIGINAL PURPOSE: (generic)

EXTRA DATA USED: (none)

SUBROUTINES CALLED:

    `param_area.aml`

DESCRIPTION:

This method will determine the area by running the `param_area.aml` method and converting the units of output to hectares. If no units are specified in the input zone map coordinate system, then no unit conversion will occur and the results will be reported using the input zone map coordinate system.

REFERENCES: (none)

param_area-km.aml

DEFINITION: Reports the area of each zone in the input zone map. Expressed in kilometers.

ORIGINAL PURPOSE: (generic)

EXTRA DATA USED: (none)

SUBROUTINES CALLED:

```
param_area.aml
```

DESCRIPTION:

This method will determine the area by running the `param_area.aml` method and converting the units of output to square kilometers. If no units are specified in the input zone map coordinate system, then no unit conversion will occur and the results will be reported using the input zone map coordinate system.

REFERENCES: (none)

param_area-meters.aml

DEFINITION: Reports the area of each zone in the input zone map. Expressed in meters.

ORIGINAL PURPOSE: (generic)

EXTRA DATA USED: (none)

SUBROUTINES CALLED:

```
param_area.aml
```

DESCRIPTION:

This method will determine the area by running the `param_area.aml` method and converting the units of output to square meters. If no units are specified in the input zone map coordinate system, then no unit conversion will occur and the results will be reported using the input zone map coordinate system.

REFERENCES: (none)

param_area-miles.aml

DEFINITION: Reports the area of each zone in the input zone map. Expressed in square miles.

ORIGINAL PURPOSE: (generic)

EXTRA DATA USED: (none)

SUBROUTINES CALLED:

    param_area.aml

DESCRIPTION:

This method will determine the area by running the param_area.aml method and converting the units of output to square miles. If no units are specified in the input zone map coordinate system, then no unit conversion will occur and the results will be reported using the input zone map coordinate system.

REFERENCES: (none)

param_area-smallest.aml

DEFINITION: Determines the identity of the smallest zone in the input zone map.

ORIGINAL PURPOSE: (generic)

EXTRA DATA USED: (none)

SUBROUTINES CALLED:

    param_area.aml

DESCRIPTION:

This method will determine the identity of the smallest zone in the input zone map. The output GRID will only contain valid values for those cells locations that correspond to those cells that constitute the smallest zone in the input zone map. The value for those cells in the output GRID will match those of the input.

This parameterization method is not envisioned as generating output for direct use in a simulation model, but as an intermediate analysis data product for use by other methods.

REFERENCES: (none)

param_area-total-nhru-acres.aml

DEFINITION: Reports the area of each zone in the input zone map. Expressed in acres.

ORIGINAL PURPOSE: `basin_area` (PRMS, replaced by `param_area-acres.aml`)

EXTRA DATA USED:

ArcInfo GRID of nhru (hydrologic response units)

SUBROUTINES CALLED:

`param_area-acres.aml`

DESCRIPTION:

The PRMS model (Leavesley and others, 1983) compares the sum of the areas of all hydrologic response units (HRUs) with the area of the entire basin (which is expressed independently as a separate parameter value). It was found that the HRU areas, derived with `param_area-acres.aml`, were frequently summing to a number slightly less than the basin area. This was largely caused by imperfections in the DEM that resulted in portions of the total basin area being "lost" when deriving the HRU map according to the user's HRU delineation methodology. In order to overcome this source of inconsistency in the produced parameters, the area for the basin was not derived by applying the `param_area-acres.aml` parameterization method to the ArcInfo GRID of the entire basin. Instead, the basin area was inferred by summing the area of the HRUs. The current method implements this alternate approach.

During the execution of the method, the user will be prompted to specify the identity of the HRU map using a GRID Browser tool, unless the identity is already known to the Parameterization Engine. The identity of the HRU map may be known if it has been specified as the input zone map (Geographic Feature Type) for other parameters being executed during the same parameterization session. If the method is being run from the **Pan/Zoom>>Command Line**, then the method will have no prior knowledge of the HRU map identity.

The output area units will be acres, unless no units are specified in the input zone map coordinate system. If no units are specified in the input zone map coordinate system, then `param_area-total-nhru-acres.aml` will report the area in the native units of the input zone map.

REFERENCES: (none)

param_area-total-nhru-km.aml

DEFINITION: Reports the area of each zone in the input zone map. Expressed in kilometers.

ORIGINAL PURPOSE: (generic)

EXTRA DATA USED:

ArcInfo GRID of nhru (hydrologic response units)

SUBROUTINES CALLED:

`param_area-km.aml`

DESCRIPTION:

This routine was implemented as a metric counterpart to the `param_area-total-nhru-acres.aml` parameterization method. See above for description.

During the execution of the method, the user will be prompted to specify the identity of the HRU map using a GRID Browser tool, unless the identity is already known. The identity of the HRU map may be known if it has been specified as the input zone map (Geographic Feature Type) for other parameters being executed during the same parameterization session. If the method is being run from the **Pan/Zoom>>Command Line**, then the method will have no prior knowledge of the HRU map identity.

The output area units will be kilometers, unless no units are specified in the input zone map coordinate system. If no units are specified in the input zone map coordinate system, then `param_area-total-nhru-km.aml` will report the area in the native units of the input zone map.

REFERENCES:

Leavesley, G.H., Lichty, R.W., Troutman, B.M., and Saindon, L.G., 1983, Precipitation-Runoff Modeling System User's Manual: U.S. Geological Survey Water-Resources Investigations Report 83-4238, p. 207.

param_aspect-reclass-dominant.aml

DEFINITION: Determines the most commonly occurring category of reclassified aspect for each zone in the input zone map.

ORIGINAL PURPOSE: (generic)

EXTRA DATA USED:

ArcInfo GRID of aspect

SUBROUTINES CALLED:

`regrouper.aml`

DESCRIPTION:

This method will determine the most commonly occurring category of reclassified aspect for each zone in the input zone map.

If no GRID of reclassified aspect exists, then the GRID of raw aspect will automatically be reclassified during the execution of the parameterization method, using the `regrouper.aml`. `param_asp-dom.aml` will then use this product, found at output/surfaces/<dem>/aspect/reclass/grid. The notation <dem> refers to the Input Elevation Grid or the filled derivative (if the user elected to fill the Input Elevation Grid).

The floating-point values found in the aspect grid range from 0-360, with 0 corresponding to north, 90 to east, 180 to south, 270 to west, and 360 (also) to north.

The default scheme for reclassifying the aspect GRID has identifiers corresponding to the center value for 45-degree ranges of values. If the user has run the parameterization method in the interactive mode and adjusted the reclassification scheme, then the output from `param_asp-dom.aml` will reflect this customized scheme. The user is cautioned that this customized scheme should be appropriate for the simulation model that will consume these results.

REFERENCES: (none)

param_basin-area-pct.aml

DEFINITION: Determines the percent that each zone in the input zone map occupies of the total basin area.

ORIGINAL PURPOSE: (generic)

EXTRA DATA USED:

ArcInfo GRID of the basin outline.

SUBROUTINES CALLED:

`param_area.aml`

DESCRIPTION:

This routine will divide the area of each zone in the input zone map by the area of another GRID. This secondary GRID is intended to represent a basin or a single-zoned zone map of the entire geographic extent of the modeling domain (that is, the AOI). The result is an expression of what percentage of the total modeling domain that an individual zone in the input zone map occupies.

The output is expressed in floating point numbers ranging from 0-1.0.

REFERENCES: (none)

param_centroid-lat.aml

DEFINITION: Determines the latitude of the geometric center for each zone in the input zone map.

ORIGINAL PURPOSE: `radpl_lat` (PRMS)

EXTRA DATA USED: (none)

SUBROUTINES CALLED:

    centroid-latlong.aml

DESCRIPTION:

This method determines the latitude of the center points for the zones in the input zone map. To do this, it relies on the `ZONALCENTROID()` function in ArcInfo GRID, which has been known to fail if zone outlines are complex or heavily crenulated. This method also relies on an automated tool to project the coordinates of the centroids into geographic coordinates. This tool has been broadly, but not exhaustively, tested for many coordinate systems used in North America. It has been known to fail for polar coordinate systems.

REFERENCES:

Leavesley, G.H., Lichty, R.W., Troutman, B.M., and Saindon, L.G., 1983, Precipitation-Runoff Modeling System - User's Manual: U.S. Geological Survey Water-Resources Investigations 83-4238, 207 p.

param_centroid-long.aml

DEFINITION: Determines the longitude of the geometric center for each zone in the input zone map.

ORIGINAL PURPOSE: (generic)

EXTRA DATA USED: (none)

SUBROUTINES CALLED:

Table 1. Reclassification scheme for converting land use/land cover to cover type.

Input land use/land cover	Output cover type
White-red-jack pine	Coniferous
Spruce-fir	Coniferous
Longleaf-slash pine	Coniferous
Loblolly-shortleaf pine	Coniferous
Oak-pine	Deciduous
Oak-hickory	Deciduous
Oak-gum-cypress	Deciduous
Elm-ash-cottonwood	Deciduous
Maple-beech-birch	Deciduous
Aspen-birch	Deciduous
Douglas-fir	Coniferous
Hemlock-Sitka spruce	Coniferous
Ponderosa pine	Coniferous
Western white pine	Coniferous
Lodgepole pine	Coniferous
Larch	Deciduous
Fir-spruce	Coniferous
Redwood	Coniferous
Chaparral	Shrub
Pinyon-juniper	Coniferous
Western hardwoods	Deciduous
Aspen-birch	Deciduous
Nonforest	Bare
Urban or Built-up Land	Bare
Dryland Cropland and Pasture	Grass
Irrigated Cropland and Pasture	Grass
Mixed Dryland/Irrigated Cropland and Pasture	Grass
Cropland/Grassland Mosaic	Grass
Cropland/Woodland Mosaic	Shrub
Grassland	Grass
Shrubland	Shrub
Mixed Shrubland/Grassland	Shrub
Chaparral	Shrub
Savanna	Grass
Broadleaf Coniferous Forest	Coniferous
Evergreen Coniferous Forest	Coniferous
Subalpine Forest	Coniferous
Mixed Forest	Deciduous
Coniferous Forest	Coniferous
Evergreen Broadleaf Forest	Coniferous
Water Bodies	Bare
Herbaceous Wetland	Shrub
Forested Wetland	Deciduous
Barren or Sparsely Vegetated	Bare
Wooded Tundra	Deciduous
Herbaceous Tundra	Shrub
Bare Ground Tundra	Grass
Wet Tundra	Grass
Mixed Tundra	Grass
Perennial Snowfields or Glaciers	Bare

Table 2. Weighting of vegetation density based on cover type.

Input cover type	Output vegetation density weight
Bare	1
Grass	15
Shrub	25
Deciduous Trees	100
Coniferous Trees	100

```
centroid-latlong.aml
```

DESCRIPTION:

This method is a counterpart to `param_centroid-lat.aml`, although the output is not used by PRMS. Please see the description above for more information on caveats associated with these methods.

REFERENCES: (none)

param_centroid-x.aml

DEFINITION: Determines the x coordinate of the geometric center for each zone in the input zone map, expressed in the coordinate system of the input zone map.

ORIGINAL PURPOSE: `hru_x` (the PRMS XYZ climate-distribution module)

EXTRA DATA USED: (none)

SUBROUTINES CALLED:

```
centroid-xy.aml
```

DESCRIPTION:

This method functions in a manner similar to `param_centroid-lat.aml`, although the output is used for different purposes. Please see previous description for more information on caveats associated with these methods.

The units of the output are presumed by the PRMS XYZ climate-distribution module to be meters, although the parameterization method does not enforce this. The user should use an Input Elevation Grid XY units that are meters if use of the PRMS XYZ climate-distribution model is intended.

REFERENCES:

Leavesley, G.H., Lichty, R.W., Troutman, B.M., and Saindon, L.G., 1983, Precipitation-Runoff Modeling System - User's Manual: U.S. Geological Survey Water-Resources Investigations 83-4238, 207 p.

param_centroid-y.aml

DEFINITION: Determines the y coordinate of the geometric center for each zone in the input zone map, expressed in the coordinate system of the input zone map.

ORIGINAL PURPOSE: `hru_y` (the PRMS XYZ climate-distribution module)

EXTRA DATA USED: (none)

SUBROUTINES CALLED:

    `centroid-xy.aml`

DESCRIPTION:

This method functions in a manner similar to param_centroid-lat.aml, although the output is used for different purposes. Please see previous description for more information on caveats associated with these methods.

The units of the output are presumed by the PRMS XYZ climate-distribution module to be meters, although the parameterization method does not enforce this. The user should use an Input Elevation Grid XY units that are meters if use of the PRMS XYZ climate-distribution model is intended.

REFERENCES:

Leavesley, G.H., Lichty, R.W., Troutman, B.M., and Saindon, L.G., 1983, Precipitation-Runoff Modeling System - User's Manual: U.S. Geological Survey Water-Resources Investigations 83-4238, 207 p.

param_chan-width.aml

DEFINITION: Allows a user to interactively set widths for each zone in the input zone map. Units are unspecified.

ORIGINAL PURPOSE: (generic)

EXTRA DATA USED:

    User input

SUBROUTINES CALLED: (none)

DESCRIPTION:

This method is intended to allow a user to interactively specify the widths of channels. Therefore, the input zone map is expected to depict some kind of drainage network (although there is no automated checking by the method to ensure this). In order to aid the user's determination of the representative width for each link in the drainage network, the method will determine mean flow accumulation for each zone and present these values to the user through a menu. The user enters values in this menu and then submits the values.

REFERENCES: (none)

param_cov-den-summer.aml

DEFINITION: Derives the mean summer vegetation cover density for each zone in the input zone map. Expressed as percent (0.00-1.00).

ORIGINAL PURPOSE: `covden_sum` (PRMS, replaced by `param_cov-den-dominant.aml`)

EXTRA DATA USED:

ArcInfo GRID of vegetation density, .../`data_bin/forests/density`

ArcInfo GRID of vegetation species, .../`data_bin/forests/lower48`

A reclassification scheme to derive a GRID of vegetation density weightings based on the `lower48` GRID (default: .../`weasel/src_aml/cov-type-wt.rmp`)

SUBROUTINES CALLED:

`sfc_cov-den-summer.aml`

DESCRIPTION:

This method will determine the mean summer vegetation density for each zone identified by a unique value in the attribute table of the input zone map. Any type of zone map, such as a MRU, subbasin, or basin map, may be used as the input to this method. Vegetation density is defined as the percentage of the ground surface that is covered if looking straight down.

This method relies on a GRID of vegetation density, .../`data_bin/forests/density`, to derive this information. This GRID should be adjusted for consumption by PRMS, as a function of the vegetation speciation found in the GRID .../`data_bin/forests/lower48`. In the noninteractive mode of execution, a default reclassification scheme is used to first simplify the vegetation speciation data into five cover types [0 = `bare`, 1 = `grass`, 2 = `shrub`, 3 = `deciduous`, 4 = `coniferous`], and from that to derive per-cover type weightings of the density. The terms shown in the Input Vegetation Species column of this table correspond to the

Table 3. Reclassification scheme for converting cover type to leaf loss.

Input cover type	Output leaf loss
Bare	100
Grass	50
Shrub	60
Deciduous Trees	60
Coniferous Trees	0

codes listed in the Data Bin section of the "Advanced Topics" section. In the interactive mode of execution, the user is presented with the default scheme and can make adjustments as they choose. The default reclassification scheme for cover type is described in table 1. The default reclassification scheme for converting cover type into a weighting is described in table 2.

As with any reclassification in the GIS Weasel, the user may substitute an alternate GRID for the default that is presented by a method. The user also should provide a corresponding alternate reclassification scheme. If an alternate reclassification scheme is to be used, then it should follow the format specification for *remap tables*, found in the ArcInfo on-line documentation. A poorly formed remap table will result in a GIS Weasel *error*.

Weighting of vegetation density by cover type is intended to properly relate percentage ground cover values for different types of cover to all other cover types. This may be important if the user feels that to properly average the percentage coverage of an area that is 50 percent covered with coniferous trees with an area that is 50 percent covered with grass, some adjustment must be made to one or both percentages.

REFERENCES:

Leavesley, G.H., Lichty, R.W., Troutman, B.M., and Saindon, L.G., 1983, Precipitation-Runoff Modeling System - User's Manual: U.S. Geological Survey Water-Resources Investigations 83-4238, 207 p.
Powell, D.S., Faulkner, J.L., Darr, D.R., Zhu, Zhiliang, and MacCleery, D.W., 1993, Forest resources of the United States, 1992, General Technical Report RM-234 (revised): Fort Collins, Colorado, U.S. Department of Agriculture, Forest Service, 132 p.

param_cov-den-summer2.aml

DEFINITION: Derives the mean summer vegetation cover density for each zone in the input zone map. Expressed as percent (0.00-1.00).

ORIGINAL PURPOSE: (generic)

EXTRA DATA USED:

ArcInfo GRID of vegetation density, .../`data_bin/forests/density`

SUBROUTINES CALLED: (none)

DESCRIPTION:

This method simply finds the mean of the vegetation density for each zone in the input zone map. No adjustments, reclassifications, or weightings are applied (as is the case for `param_cov-den-summer.aml`). This routine was developed for users that may have access to a higher quality map of vegetation density than was available at the time of the development of the `param_cov-den-summer.aml` parameterization method.

REFERENCES: (none)

param_cov-den-summer-dominant.aml

DEFINITION: Determines the most commonly occurring vegetation species in a zone and then derives the mean vegetation density for that type, and assigns that value to the respective zone in in the input zone map. Expressed as percent (0.00-1.00).

ORIGINAL PURPOSE: `covden_sum` (PRMS)

EXTRA DATA USED:

ArcInfo GRID of vegetation density, …`/data_bin/forests/density`

ArcInfo GRID of vegetation species, …`/data_bin/forests/lower48`

ArcInfo GRID of elevation, referenced by variable `%.aGrid%`

SUBROUTINES CALLED:

`timber-line.aml`

`param_elevation-median-feet.aml`

`param_tree-dom.aml`

DESCRIPTION:

This method replaces `param_cov-den-summer.aml` for use with PRMS.

This routine first determines the dominant cover type for each zone in the input zone map. It will then assign the average vegetation density for this cover type to the entirety of the surrounding zone (of the input zone map). If, at this point, the median elevation for the zone (of the input zone map) is found to exceed timber line (by default, set to 11,500 feet), a vegetation density of 0 is forced.

REFERENCES:

Leavesley, G.H., Lichty, R.W., Troutman, B.M., and Saindon, L.G., 1983, Precipitation-Runoff Modeling System - User's Manual: U.S. Geological Survey Water-Resources Investigations 83-4238, 207 p.
Powell, D.S., Faulkner, J.L., Darr, D.R., Zhu, Zhiliang, and MacCleery, D.W., 1993, Forest resources of the United States, 1992, General Technical Report RM-234 (revised): Fort Collins, Colorado, U.S. Department of Agriculture, Forest Service, 132 p.

param_cov-den-winter.aml

DEFINITION: Derives the mean winter vegetation cover density for each zone in the input zone map. Expressed as percent (0.00-1.00).

ORIGINAL PURPOSE: covden_win (PRMS, replaced by param_cov-den-winter3.aml)

EXTRA DATA USED:

ArcInfo GRID of vegetation density, .../data_bin/forests/density

ArcInfo GRID of vegetation species, .../data_bin/forests/lower48

SUBROUTINES CALLED:

sfc_cov-den-winter.aml

sfc_leaf-loss.aml

DESCRIPTION:

This method will determine the mean winter vegetation cover density for each zone in the input zone map.

Winter vegetation density is derived for every cell by applying a leaf-loss factor to the summer vegetation density. The reader is referred to the description for the param_cov-den-summer.aml for more details on the calculation of the summer vegetation density. The leaf-loss factor is derived in two-step process. First the GRID of vegetation species is reclassified into a GRID of cover type using the same scheme as shown in table 1. Then the derived cover type GRID is reclassified into a leaf-loss factor using the reclassification scheme shown in table 3.

The winter vegetation density surface (that is, on a per-cell basis) is calculated by subtracting the expected proportion of winter reduction in foliage for each cover type from 1.0 and multiplying the result by the summer vegetation density.

If the user chooses to run this method in the interactive mode, then alternate GRIDs of vegetation density and speciation can be used, as well as alternate reclassification schemes. If an alternate reclassification scheme is to be used, then it should follow the format specification for remap tables, found in the ArcInfo on-line documentation. A poorly formed REMAP TABLE will result in an error in the execution of the method. Once the mean winter vegetation density is calculated within the interactive mode, the user is presented with an editable menu of these values. This functionality exists to allow the user to modify the values prior to the generation of parameter files.

REFERENCES:

Leavesley, G.H., Lichty, R.W., Troutman, B.M., and Saindon, L.G., 1983, Precipitation-Runoff Modeling System - User's Manual: U.S. Geological Survey Water-Resources Investigations 83-4238, 207 p.
Powell, D.S., Faulkner, J.L., Darr, D.R., Zhu, Zhiliang, and MacCleery, D.W., 1993, Forest resources of the United States, 1992, General Technical Report RM-234 (revised): Fort Collins, Colorado, U.S. Department of Agriculture, Forest Service, 132 p.

param_cov-den-winter2.aml

DEFINITION: Derives the mean winter vegetation cover density for each zone in the input GRID of polygon-type features, expressed as percent (0.00-1.00).

ORIGINAL PURPOSE: `covden_win` (PRMS, replaced by `param_cov-den-winter3.aml`)

EXTRA DATA USED:

ArcInfo GRID of vegetation density, …/`data_bin/forests/density`

ArcInfo GRID of vegetation species, …/`data_bin/forests/lower48`

SUBROUTINES CALLED:

`sfc_cov-type-prms.aml`

`param_cov-type-prms3.aml`

DESCRIPTION:

This method has been replaced by `param_cov-den-winter3.aml`, which is more accurate.

First the per-cell PRMS cover type is determined (please see the description for param_cov-type-prms.aml for more detail). PRMS cover type follows the scheme of [`0=bare, 1=grass, 2=shrub, 3=tree`]. Then the per-zone cover type is determined, using `param_cov-type-prms3.aml` (described later). Where corresponding cells in these two layers share cover types, the values from a third GRID, …/`data_bin/forests/density`, are selected into a temporary GRID of summer vegetation density. This operation has the effect of eliminating density values that are not associated with the desired cover type within a zone of the input zone map.

A leaf "keep" factor (as opposed to a leaf-loss factor) is set based on the PRMS cover type. The leaf-keep factors, [`0, 0.8, 0.7. 1.00`], correspond to the PRMS cover types [`0=bare, 1=grass, 2=shrub, 3=tree`], respectively. The leaf-keep factor is then multiplied by the temporary summer vegetation density. This product is then averaged for each zone in the input zone map to derive the winter vegetation density, expressed as a number between 0-100.

REFERENCES:

Leavesley, G.H., Lichty, R.W., Troutman, B.M., and Saindon, L.G., 1983, Precipitation-Runoff Modeling System - User's Manual: U.S. Geological Survey Water-Resources Investigations 83-4238, 207 p.

Powell, D.S., Faulkner, J.L., Darr, D.R., Zhu, Zhiliang, and MacCleery, D.W., 1993, Forest resources of the United States, 1992, General Technical Report RM-234 (revised): Fort Collins, Colorado, U.S. Department of Agriculture, Forest Service, 132 p.

param_cov-den-winter3.aml

DEFINITION: Derives the mean winter vegetation cover density for each zone in the input zone map. Expressed as percent (0.00-1.00).

ORIGINAL PURPOSE: `covden_win` (PRMS)

EXTRA DATA USED:

ArcInfo GRID of vegetation density, …/`data_bin/forests/density`

ArcInfo GRID of vegetation species, …/`data_bin/forests/lower48`

SUBROUTINES CALLED:

`sfc_cov-type-prms.aml`

`param_cov-type-prms3.aml`

DESCRIPTION:

This method is almost identical to `param_cov-den-winter2.aml`. The main difference is that the mathematics of the newer version have been adjusted to more accurately maintain floating-point numerical operations throughout the method. The reader is referred to the description of param_cov-den-winter2.aml for more detail.

REFERENCES: (none)

param_cov-type.aml

DEFINITION: Determines the most commonly occurring cover type for each zone in the input zone map. Expressed (0=Bare, 1=Grass, 2=Shrub, 3=Deciduous Trees, 4=Coniferous Trees).

Table 4.   Reclassification scheme for converting land use/land cover to PRMS cover type.

Input land use/land cover	Output PRMS cover type
White-red-jack pine	Trees
Spruce-fir	Trees
Longleaf-slash pine	Trees
Loblolly-shortleaf pine	Trees
Oak-pine	Trees
Oak-hickory	Trees
Oak-gum-cypress	Trees
Elm-ash-cottonwood	Trees
Maple-beech-birch	Trees
Aspen-birch	Trees
Douglas-fir	Trees
Hemlock-Sitka spruce	Trees
Ponderosa pine	Trees
Western white pine	Trees
Lodgepole pine	Trees
Larch	Trees
Fir-spruce	Trees
Redwood	Trees
Chaparral	Shrub
Pinyon-juniper	Trees
Western hardwoods	Trees
Aspen-birch	Trees
Nonforest	Bare
Urban or Built-up Land	Bare
Dryland Cropland and Pasture	Grass
Irrigated Cropland and Pasture	Grass
Mixed Dryland/Irrigated Cropland and Pasture	Grass
Cropland/Grassland Mosaic	Grass
Cropland/Woodland Mosaic	Grass
Grassland	Grass
Shrubland	Shrub
Mixed Shrubland/Grassland	Shrub
Chaparral	Shrub
Savanna	Grass
Broadleaf Deciduous Forest	Trees
Evergreen Coniferous Forest	Trees
Subalpine Forest	Trees
Mixed Forest	Trees
Deciduous Coniferous Forest	Trees
Evergreen Broadleaf Forest	Trees
Water Bodies	Bare
Herbaceous Wetland	Shrub
Forested Wetland	Trees
Barren or Sparsely Vegetated	Bare
Wooded Tundra	Trees
Herbaceous Tundra	Shrub
Bare Ground Tundra	Bare
Wet Tundra	Bare
Mixed Tundra	Grass
Perennial Snowfields or Glaciers	Bare

ORIGINAL PURPOSE: (generic)

EXTRA DATA USED:

ArcInfo GRID of land use/land cover, …/data_bin/lcov_comp/lulc

SUBROUTINES CALLED:

sfc_cov-type.aml

DESCRIPTION:

This method will determine the most commonly occurring cover type for each zone in the input zone map.

The default reclassification of land use/land cover into cover type is defined according to table 1. The terms shown in the Output Cover Type column of this table correspond to the code scheme [bare=0, grass=1, shrub=2, deciduous=3, coniferous=4]. If this method is run noninteractively, then this default scheme is used.

As with any reclassification in the GIS Weasel, the user may substitute an alternate GRID for the default that is presented by a method. The user also should provide a corresponding alternate reclassification scheme. If an alternate reclassification scheme is to be used, then it should follow the format specification for remap tables, found in the ArcInfo on-line documentation. A poorly formed remap table will result in a GIS Weasel error.

This method is used by a variety of other parameterization methods, notably those for calculating cover density.

REFERENCES:

Powell, D.S., Faulkner, J.L., Darr, D.R., Zhu, Zhiliang, and MacCleery, D.W., 1993, Forest resources of the United States, 1992, General Technical Report RM-234 (revised): Fort Collins, Colorado, U.S. Department of Agriculture, Forest Service, 132 p.

param_cov-type-prms.aml

DEFINITION: Determines the most commonly occurring PRMS cover type for each zone in the input zone map, expressed (0=Bare, 1=Grass, 2=Shrub, 3=Trees).

ORIGINAL PURPOSE: cov_type (PRMS, replaced by param_cov-type-prms3.aml)

EXTRA DATA USED:

ArcInfo GRID of land use/land cover, …/`data_bin/lcov_comp/lulc`

SUBROUTINES CALLED:

`sfc_cov-type-prms.aml`

DESCRIPTION:

This method will determine the most commonly occurring PRMS cover type for each zone in the input zone map. The PRMS cover type scheme does not discriminate between deciduous and coniferous trees, as the standard cover type scheme (shown in table 1).

The default reclassification of land use/land cover into cover type is defined according to table 4. The terms shown in the output PRMS Cover Type column of this table correspond to the code scheme `[bare=0, grass=1, shrub=2, trees=3]`. If this method is run noninteractively, then this default scheme is used.

As with any reclassification in the GIS Weasel, the user may substitute an alternate GRID for the default that is presented by a method. The user also should provide a corresponding alternate reclassification scheme. If an alternate reclassification scheme is to be used, then it should follow the format specification for remap tables, found in the ArcInfo on-line documentation. A poorly formed remap table will result in a GIS Weasel error.

REFERENCES:

Anderson, J.R., Hardy, E.E., Roach, J.T., and Witmer, R.E., 1976, A land use and land cover classification system for use with remote sensor data: U.S. Geological Survey Professional Paper 964, 28 p.
Leavesley, G.H., Lichty, R.W., Troutman, B.M., and Saindon, L.G., 1983, Precipitation-Runoff Modeling System - User's Manual: U.S. Geological Survey Water-Resources Investigations 83-4238, 207 p.
Powell, D.S., Faulkner, J.L., Darr, D.R., Zhu, Zhiliang, and MacCleery, D.W., 1993, Forest resources of the United States, 1992, General Technical Report RM-234 (revised): Fort Collins, Colorado, U.S. Department of Agriculture, Forest Service, 132 p.

param_cov-type-prms2.aml

DEFINITION: Determines the cover type for each zone in the input zone map. Expressed (0=Bare, 1=Grass, 2=Shrub, 3=Trees).

ORIGINAL PURPOSE: `cov_type` (PRMS, replaced by `param_cov-type-prms3.aml`)

EXTRA DATA USED:

ArcInfo GRID of land use/land cover, …/`data_bin/lcov_comp/lulc`

SUBROUTINES CALLED:

`sfc_cov-type-prms.aml`

DESCRIPTION:

This method is a more complex approach than that implemented by `param_cov-type-prms.aml`. It was implemented in order to not merely describe the most commonly occurring cover type but to describe the most hydrologically important cover type (as far as the PRMS model is concerned), even if this cover type did not occupy the largest proportion of the zones in the input zone map.

First, this method determines on a per-cell basis the PRMS cover type using the scheme shown in table 4. Then it evaluates this surface on a per-zone basis, which is the most important cover type. More specifically, if the zone is more than 80 percent bare, then this is the output cover type. If not, then if the zone is more than 20 percent tree, then it is described as tree. If neither of these conditions exists, and if shrubs occupy more than 20 percent, the output cover type is shrub. Failing this, if the combined percent of trees and shrubs exceeds 35 percent, then the more common of the two categories is assigned to the output. If this condition is not met, and if grass occupies more than 50 percent, then the output is assigned a grass cover type. If this last condition does not occur, then the largest nonbare category is selected.

REFERENCES:

Anderson, J.R., Hardy, E.E., Roach, J.T., and Witmer, R.E., 1976, A land use and land cover classification system for use with remote sensor data: U.S. Geological Survey Professional Paper 964, 28 p.

Leavesley, G.H., Lichty, R.W., Troutman, B.M., and Saindon, L.G., 1983, Precipitation-Runoff Modeling System - User's Manual: U.S. Geological Survey Water-Resources Investigations 83-4238, 207 p.

Powell, D.S., Faulkner, J.L., Darr, D.R., Zhu, Zhiliang, and MacCleery, D.W., 1993, Forest resources of the United States, 1992, General Technical Report RM-234 (revised): Fort Collins, Colorado, U.S. Department of Agriculture, Forest Service, 132 p.

param_cov-type-prms3.aml

DEFINITION: Determines the cover type for each zone in the input zone map. Expressed (0=Bare, 1=Grass, 2=Shrub, 3=Trees).

ORIGINAL PURPOSE: `cov_type` (PRMS)

EXTRA DATA USED:

ArcInfo GRID of land use/land cover, .../`data_bin/lcov_comp/lulc`

SUBROUTINES CALLED:

`sfc_cov-type-prms.aml`

DESCRIPTION:

This method is a minor evolution of that implemented by `param_cov-type-prms2.aml`. Please see the description for `param_cov-type-prms2.aml` for more detail. The main difference is that the initial conditional threshold of 80 percent bare being assigned an output cover type of "bare" was raised to 90 percent.

This routine does not represent a methodological advance, but a fine tuning of the approach to compensate for characteristics of the GRID of land use/land cover.

REFERENCES:

Anderson, J.R., Hardy, E.E., Roach, J.T., and Witmer, R.E., 1976, A land use and land cover classification system for use with remote sensor data: U.S. Geological Survey Professional Paper 964, 28 p.

Leavesley, G.H., Lichty, R.W., Troutman, B.M., and Saindon, L.G., 1983, Precipitation-Runoff Modeling System - User's Manual: U.S. Geological Survey Water-Resources Investigations 83-4238, 207 p.

Powell, D.S., Faulkner, J.L., Darr, D.R., Zhu, Zhiliang, and MacCleery, D.W., 1993, Forest resources of the United States, 1992, General Technical Report RM-234 (revised): Fort Collins, Colorado, U.S. Department of Agriculture, Forest Service, 132 p.

param_daf_pct_area.aml

DEFINITION: Determine the percentage of the contributing area associated with the zone in the input zone map that drains to that zone. Expressed as percent.

ORIGINAL PURPOSE: `daf_pct_area` (DAFLOW)

EXTRA DATA USED:

ArcInfo GRID of stream links, referenced as `%ndabranch%`

SUBROUTINES CALLED:

`zone_one-plane.aml`

DESCRIPTION:

The input zone map is expected to depict the *segments* of a drainage network, referred to as *branches*, associated with the application of the DAFLOW model. The segments can be thought of as points along a link in a branch network. This parameterization method will determine the percentage of the contributing area associated with this segment that drains into that segment.

During the execution of the method, the user is prompted to specify the identity of the map of drainage links (*ndabranch*). The `zone_one-plane.aml` routine is used to derive the contributing areas to each link in ndabranch. The `zone_one-plane.aml` routine is then reused to derive the contributing areas to each segment (zone) in the input zone map. The number of cells in the contributing area of a segment is divided by the number of cells in the corresponding ndabranch contributing area to determine a percentage.

The segment map is usually created using the **Tool Panel>>Automated Tools>>DAFLOW-type Zones>>Nodes** tool.

REFERENCES:

Jobson, H.E., 1989, Users manual for an open-channel streamflow model based on the diffusion analogy: U.S. Geological Survey Water-Resources Investigations Report 89-4133, 73 p.

param_dajunction-down.aml

DEFINITION: Returns the identity of the junction at the downstream end of each zone in the input zone map.

ORIGINAL PURPOSE: `daf_jncd` (DAFLOW)

EXTRA DATA USED:

ArcInfo GRID of stream junctions, referenced as `%ndajunction%`

SUBROUTINES CALLED:

`zone_one-plane.aml`

DESCRIPTION:

The input zone map is expected to represent the links of a drainage network associated with the application of the DAFLOW model. The features in the map are called *branches* in this model. During the execution of the method, the user is prompted to specify the identity of the map of stream junctions (*ndajunction*). The junctions are assumed to fall on some part of a branch. The method will then determine the downstream-most cell of each branch and record the identity of the junction that is found at this cell.

The junction map is usually created using the **Tool Panel>>Automated Tools>>DAFLOW-type Zones>>Junctions** tool in order to ensure the proper location of the junctions on the branches.

REFERENCES:

Jobson, H.E., 1989, Users manual for an open-channel streamflow model based on the diffusion analogy: U.S. Geological Survey Water-Resources Investigations Report 89-4133, 73 p.

param_dist2headwater.aml

DEFINITION: Determines the minimum hydrologic distance from each zone in the input zone map to the beginning of an associated zone in a GRID of drainage links. The input zone map is assumed to be DAFLOW stream segments. Expressed in the XY units of the input zone map.

ORIGINAL PURPOSE: (generic)

EXTRA DATA USED:

ArcInfo GRID of a drainage network, referenced as `%ndabranch%`

SUBROUTINES CALLED:

`zone_flowlength-up.aml`

DESCRIPTION:

The input zone map is assumed to be DAFLOW stream segments, which are points on the links of a drainage network. This map is typically created using the **Tool Panel>>Automated Methods>>DAFLOW-type Zones>>Nodes t**ool.

During the execution of the method, the user is prompted to specify the identity of the drainage network, referred to as the *branches* in DAFLOW, from which the input zone map of stream segments was derived. A map of the hydrologic distance from the headwater point (that is, the upstream most cell) of each link in the map of branches to the cells in the zones of the input zone map is derived on a per-cell basis. The smallest distance within each zone of the input zone map is reported.

REFERENCES:

Jobson, H.E., 1989, Users manual for an open-channel streamflow model based on the diffusion analogy: U.S. Geological Survey Water-Resources Investigations Report 89-4133, 73 p.

param_dist2headwater-miles.aml

DEFINITION: Determine the minimum hydrologic distance from each zone in the input zone map to the beginning of an associated zone in a GRID of drainage links. The input zone map is assumed to be DAFLOW stream segments. Expressed in miles.

ORIGINAL PURPOSE: `daf_x` (DAFLOW)

EXTRA DATA USED: (none)

SUBROUTINES CALLED:

`param_dist2headwater.aml`

DESCRIPTION:

This parameterization method runs `param_dist2headwater.aml` and converts the units of the output to miles. Please see the description of `param_dist2headwater.aml` for more information.

REFERENCES:

Jobson, H.E., 1989, Users manual for an open-channel streamflow model based on the diffusion analogy: U.S. Geological Survey Water-Resources Investigations Report 89-4133, 73 p.

param_down-id.aml

DEFINITION: Determines the identity of the downstream zone for each zone in the input zone map.

ORIGINAL PURPOSE: (generic)

EXTRA DATA USED: (none)

SUBROUTINES CALLED:

`zone_outlet-downstream.aml`

`sfc_flow-direction.aml`

DESCRIPTION:

Subroutine `zone_outlet-downstream.aml` finds for each zone in the input zone map the cell with the maximum flow accumulation. These cells are extracted into a new GRID that can be thought of as the outlets of the zones. Subroutine `param_down-id.aml` then reports the value of the cell (of the input zone map) pointed to by the flow direction of each outlet cell.

This method is not currently in use. Subroutine `zone_outlet-downstream.aml` has been replaced by a much faster implementation of subroutine `zone_outlet-downstream2.aml`.

REFERENCES: (none)

param_elevation-max-meters.aml

DEFINITION: Determines the maximum elevation for each zone in the input zone map. Expressed in meters.

ORIGINAL PURPOSE: (generic)

EXTRA DATA USED:

ArcInfo GRID of elevation, referenced as `%.aGrid%`

SUBROUTINES CALLED: (none)

DESCRIPTION: (see definition)

REFERENCES: (none)

param_elevation-mean-feet.aml

DEFINITION: Determines the mean elevation for each zone in the input zone map. Expressed in feet.

ORIGINAL PURPOSE: (generic)

EXTRA DATA USED:

ArcInfo GRID of elevation, referenced as `%.aGrid%`

SUBROUTINES CALLED: (none)

DESCRIPTION: (see definition)

REFERENCES: (none)

param_elevation-mean-meters.aml

DEFINITION: Determines the mean elevation for each zone in the input zone map. Expressed in meters.

ORIGINAL PURPOSE: (generic)

EXTRA DATA USED:

ArcInfo GRID of elevation, referenced as `%.aGrid%`

SUBROUTINES CALLED: (none)

DESCRIPTION: (see definition)

REFERENCES: (none)

param_elevation-median-feet.aml

DEFINITION: Determines the median elevation for each zone in the input zone map. Expressed in feet.

ORIGINAL PURPOSE: `hru_elev` (PRMS)

EXTRA DATA USED:

ArcInfo GRID of elevation, referenced as `%.aGrid%`

SUBROUTINES CALLED: (none)

DESCRIPTION: (see definition)

REFERENCES:

Leavesley, G.H., Lichty, R.W., Troutman, B.M., and Saindon, L.G., 1983, Precipitation-Runoff Modeling System - User's Manual: U.S. Geological Survey Water-Resources Investigations 83-4238, 207 p.

param_elevation-median-meters.aml

DEFINITION: Determines the median elevation for each zone in the input zone map. Expressed in meters.

ORIGINAL PURPOSE: (generic)

EXTRA DATA USED:

ArcInfo GRID of elevation, referenced as `%.aGrid%`

SUBROUTINES CALLED: (none)

DESCRIPTION: (see definition)

REFERENCES: (none)

param_elevation-median.aml

DEFINITION: Determines the median elevation for each zone in the input zone map. Expressed in the Z units of the elevation GRID.

ORIGINAL PURPOSE: (generic)

EXTRA DATA USED:

ArcInfo GRID of elevation, referenced as `%.aGrid%`

SUBROUTINES CALLED: (none)

DESCRIPTION: (see definition)

REFERENCES: (none)

param_elevation-min-meters.aml

DEFINITION: Derives the minimum elevation for each zone in the input zone map. Expressed in meters.

ORIGINAL PURPOSE: (generic)

EXTRA DATA USED:

ArcInfo GRID of elevation, referenced as `%.aGrid%`

SUBROUTINES CALLED: (none)

DESCRIPTION:

This method will determine the minimum elevation for each zone identified by a unique value in the input zone map value attribute table. Any type of zone map, such as a MRU, subbasin, or basin map, may be used as the input to this method.

REFERENCES: (none)

param_elevation-range-feet.aml

DEFINITION: Derives the range in elevation, expressed in feet, within each zone of the input zone map.

ORIGINAL PURPOSE: (generic)

EXTRA DATA USED:

ArcInfo GRID of elevation, referenced as `%.aGrid%`

SUBROUTINES CALLED: (none)

DESCRIPTION: (see definition)

REFERENCES: (none)

param_elevation-range-meters.aml

DEFINITION: Derives the range in elevation, expressed in feet, within each zone of the input zone map.

ORIGINAL PURPOSE: (generic)

EXTRA DATA USED:

ArcInfo GRID of elevation, referenced as `%.aGrid%`

SUBROUTINES CALLED: (none)

DESCRIPTION: (see definition)

REFERENCES: (none)

param_elevation-std-meters.aml

DEFINITION: Derives the standard deviation in elevation, expressed in feet, within each zone of the input zone map.

ORIGINAL PURPOSE: (generic)

EXTRA DATA USED:

ArcInfo GRID of elevation, referenced as `%.aGrid%`

SUBROUTINES CALLED: (none)

DESCRIPTION: (see definition)

REFERENCES: (none)

param_fac-local.aml

DEFINITION: Reports the number of cells within each zone in the input zone map.

ORIGINAL PURPOSE: (generic)

EXTRA DATA USED: (none)

SUBROUTINES CALLED:

    `param_area.aml`

DESCRIPTION:

This method will determine the number of cells for each zone in the input zone map. This method is largely unused because the same value can be more easily calculated by the syntax: `<out_grid> = zonalmax(<in_grid>, <in_grid>.count)`.

REFERENCES: (none)

param_fac-max.aml

DEFINITION: Returns the largest value of flow accumulation, expressed as the number of cells, that exists within each zone in the input zone map.

ORIGINAL PURPOSE: (generic)

EXTRA DATA USED:

    ArcInfo GRID of flow accumulation, referred to as `%.gridfac%`

SUBROUTINES CALLED:

    `sfc_flow-accumulation.aml`

DESCRIPTION:

(see definition).

REFERENCES: (none)

param_fac-max-meters.aml

DEFINITION: Returns the largest value of flow accumulation, expressed in square meters, that exists within each zone in the input zone map.

ORIGINAL PURPOSE: (generic)

EXTRA DATA USED: (none)

SUBROUTINES CALLED:

param_fac-max.aml

DESCRIPTION:

This routine simply runs the `param_fac-max.aml` method and converts those results to the equivalent value in square meters, using the cell size of the input zone map. Please see the description for `param_fac-max.aml`.

REFERENCES: (none)

param_fac-mean.aml

DEFINITION: Returns the mean value of flow accumulation, expressed as the number of cells that exist within each zone in the input zone map.

ORIGINAL PURPOSE: (generic)

EXTRA DATA USED:

ArcInfo GRID of flow accumulation, referred to as `%.gridfac%`

SUBROUTINES CALLED:

sfc_flow-accumulation.aml

DESCRIPTION:

(see definition).

REFERENCES: (none)

param_fac-min.aml

DEFINITION: Returns the minimum value of flow accumulation, expressed as the number of cells, that exists within each zone in the input zone map.

ORIGINAL PURPOSE: (generic)

EXTRA DATA USED:

ArcInfo GRID of flow accumulation, referred to as `%.gridfac%`

SUBROUTINES CALLED:

`sfc_flow-accumulation.aml`

DESCRIPTION:

(see definition).

REFERENCES: (none)

param_fac-min-meters.aml

DEFINITION: Returns the minimum value of flow accumulation, expressed in square meters, that exists within each zone in the input zone map.

ORIGINAL PURPOSE: (generic)

EXTRA DATA USED: (none)

SUBROUTINES CALLED:

`param_fac-max.aml`

DESCRIPTION:

This routine simply runs the `param_fac-min.aml` method and converts those results to the equivalent value in square meters, using the cell size of the input zone map.

REFERENCES: (none)

param_fac-pct.aml

DEFINITION: (not in use) Derives the difference between the mean and minimum flow accumulation for each zone in the input zone map, divided by the range in flow accumulation for the same zone.

ORIGINAL PURPOSE: (generic)

EXTRA DATA USED:

ArcInfo GRID of flow accumulation, referred to as `%.gridfac%`

SUBROUTINES CALLED:

```
sfc_flow-accumulation.aml

param_min.aml

param_range.aml

param_range.aml
```

DESCRIPTION:

This method will determine a measure of variation in flow accumulation that exists within each zone identified by a unique value in the input zone map value attribute table. Any type of zone map, such as a MRU, subbasin, or basin map, may be used as the input to this method.

This method is not currently in use.

REFERENCES: (none)

param_flowlength.aml

DEFINITION: For each zone in the input zone map, this method returns the *range* of the flow length between each cell in a zone and the zone outlet. Expressed in the XY units of the coordinate system of the input zone map.

ORIGINAL PURPOSE: (generic)

EXTRA DATA USED: (none)

SUBROUTINES CALLED:

```
sfc_flowlength-down.aml
```

```
param_range.aml
```

DESCRIPTION:

Flow length is defined here as the length of the path between a cell and the outlet of the zone. The path is derived according to the flow direction surface.

REFERENCES: (none)

param_flowlength-down-to-stream.aml

DEFINITION: For each zone in the input zone map, this method returns the *range* of the flow length between each cell in a zone and the nearest link in a user-supplied GRID that represents a drainage network. Expressed in the XY units of the coordinate system of the input zone map.

ORIGINAL PURPOSE: (generic)

EXTRA DATA USED:

ArcInfo GRID of a drainage network, user-specified during execution.

SUBROUTINES CALLED:

```
sfc_flowlength-down.aml
```

```
param_range.aml
```

DESCRIPTION:

Flow length is defined here as the length of the path between a cell and any cell in the user-supplied GRID (that is expected to depict a drainage network). The path is derived according to the Flow Direction surface.

REFERENCES: (none)

param_flowlength-feet.aml

DEFINITION: For each zone in the input zone map, this method returns the *range* of the flow length between each cell in a zone and the zone outlet. Expressed in feet.

ORIGINAL PURPOSE: (generic)

EXTRA DATA USED: (none)

SUBROUTINES CALLED:

    param_flowlength.aml

DESCRIPTION:

This method runs the `param_flowlength.aml` method and converts the units of the results into feet. Please see the description of `param_flowlength.aml` for more information.

REFERENCES: (none)

param_flowlength-max.aml

DEFINITION: For each zone in the input zone map, this method returns the *maximum* flow length between each cell in a zone and the zone outlet. Expressed in the XY units of the input zone map.

ORIGINAL PURPOSE: (generic)

EXTRA DATA USED: (none)

SUBROUTINES CALLED:

    sfc_flowlength-down.aml

DESCRIPTION:

Please see the description of `param_flowlength.aml` for more information on flow lengths.

REFERENCES: (none)

param_flowlength-max-meters.aml

DEFINITION: The returned value is the *maximum* flow distance in each zone in the input zone map between each cell and the zone outlet. Expressed in meters.

ORIGINAL PURPOSE: (generic)

EXTRA DATA USED: (none)

SUBROUTINES CALLED:

   `param_flowlength-max.aml`

DESCRIPTION:

This method runs the `param_flowlength-max.aml` method and converts the units of the results into meters. Please see the description of `param_flowlength-max.aml` for more information. Also, please see the description of `param_flowlength.aml` for more information on flow lengths.

REFERENCES: (none)

param_flowlength-mean.aml

DEFINITION: For each zone in the input zone map, this method returns the *mean* flow length between each cell in a zone and the zone outlet. Expressed in the XY units of the input zone map.

ORIGINAL PURPOSE: (generic)

EXTRA DATA USED: (none)

SUBROUTINES CALLED:

   `sfc_flowlength-down.aml`

DESCRIPTION:

Flow length is defined here as the length of the path between a cell and the outlet of the zone. The path is derived according to the Flow Direction surface. Also, please see the description of `param_flowlength.aml` for more information on flow lengths.

REFERENCES: (none)

param_flowlength-mean-meters.aml

DEFINITION: For each zone in the input zone map, this method returns the *mean* flow length between each cell in a zone and the zone outlet. Expressed in meters.

ORIGINAL PURPOSE: (generic)

EXTRA DATA USED: (none)

SUBROUTINES CALLED:

```
param_flowlength-mean.aml
```

DESCRIPTION:

This method runs the `param_flowlength-mean.aml` method and converts the units of the results into meters. Please see the description of `param_flowlength-mean.aml` for more information. Also, please see the description of `param_flowlength.aml` for more information on flow lengths.

REFERENCES: (none)

param_flowlength-meters.aml

DEFINITION: For each zone in the input zone map, this method returns the *range* in flow length between each cell in a zone and the zone outlet. Expressed in the meters.

ORIGINAL PURPOSE: (generic)

EXTRA DATA USED: (none)

SUBROUTINES CALLED:

```
param_flowlength.aml
```

DESCRIPTION:

This method runs the `param_flowlength.aml` method and converts the units of the results into meters. Please see the description of `param_flowlength.aml` for more information.

REFERENCES: (none)

param_flowlength-percent.aml

DEFINITION: For each zone in the input zone map, this method returns the *difference between the mean and minimum* flow length between each cell in a zone and the zone outlet, divided by the range in flow length for the same zone.

ORIGINAL PURPOSE: for use in the delineation of channel-routing increment for TOPMODEL.

EXTRA DATA USED: (none)

SUBROUTINES CALLED:

```
sfc_flowlength-down.aml
```

```
param_range.aml
```

```
param_min.aml
```

DESCRIPTION:

This method will determine a measure of variation in flow length that exists within each zone in the input zone map. Also, please see the description of `param_flowlength.aml` for more information on flow lengths.

REFERENCES:

Beven, K. J., Quinn, P.F., Romanowicz, R.J., Freer, James, Fisher, J.I., and Lamb, R., 1995, TOPMODEL and GRIDATB – A Users Guide to the Distribution Versions (95.02): Lancaster University, United Kingdom, Centre for Research on Environmental Systems and Statistics, Institute of Environmental and Biological Sciences, 31 p.

param_flowlength-up.aml

DEFINITION: Returns the *range* in the flow distances in between each cell in the zones of an input zone map and drainage divide for that cell. Expressed in the XY units of the coordinate system of the input zone map.

ORIGINAL PURPOSE: (generic)

EXTRA DATA USED: (none)

SUBROUTINES CALLED:

```
sfc_flowlength-up.aml
```

```
param_min.aml
```

```
param_max.aml
```

DESCRIPTION:

Flow length is defined here as the length of the path between a cell, termed here as a *destination*, and the nearest cell that is part of the drainage divide that feeds the destination, termed the *source*. The path is derived according to the Flow Direction surface.

REFERENCES: (none)

param_focalvariety-data.aml

DEFINITION: Determines number of distinct values surrounding each cell within the input zone map. This is frequently used to detect cells at the boundaries of cell map.

ORIGINAL PURPOSE: (generic)

EXTRA DATA USED: (none)

SUBROUTINES CALLED: (none)

DESCRIPTION:

This is frequently used to detect cells at the boundaries of cell map. The tally of the distinct values does not include cells with null (also referred to as NODATA) values. This excludes cells that are adjacent to the edge of the entire AOI. Also, see the description for `param_focalvariety-nodata.aml`.

REFERENCES: (none)

param_focalvariety-nodata.aml

DEFINITION: Determines the number of distinct values surrounding each cell within the input zone map.

ORIGINAL PURPOSE: (generic)

EXTRA DATA USED: (none)

Table 5. Reclassification scheme for converting land cover into degree of imperviousness.

Input land use/land cover	Output percent impervious
White-red-jack pine	0.00
Spruce-fir	0.00
Longleaf-slash pine	0.00
Loblolly-shortleaf pine	0.00
Oak-pine	0.00
Oak-hickory	0.00
Oak-gum-cypress	0.00
Elm-ash-cottonwood	0.00
Maple-beech-birch	0.00
Aspen-birch	0.00
Douglas-fir	0.00
Hemlock-Sitka spruce	0.00
Ponderosa pine	0.00
Western white pine	0.00
Lodgepole pine	0.00
Larch	0.00
Fir-spruce	0.00
Redwood	0.00
Chaparral	0.00
Pinyon-juniper	0.00
Western hardwoods	0.00
Aspen-birch	0.00
Nonforest	0.00
Urban or Built-up Land	1.00
Dryland Cropland and Pasture	0.00
Irrigated Cropland and Pasture	0.00
Mixed Dryland/Irrigated Cropland and Pasture	0.00
Cropland/Grassland Mosaic	0.00
Cropland/Woodland Mosaic	0.00
Grassland	0.00
Shrubland	0.00
Mixed Shrubland/Grassland	0.00
Chaparral	0.00
Savanna	0.00
Broadleaf Deciduous Forest	0.00
Evergreen Coniferous Forest	0.00
Subalpine Forest	0.00
Mixed Forest	0.00
Deciduous Coniferous Forest	0.00
Evergreen Broadleaf Forest	0.00
Water Bodies	1.00
Herbaceous Wetland	0.00
Forested Wetland	0.00
Barren or Sparsely Vegetated	0.00
Wooded Tundra	0.00
Herbaceous Tundra	0.00
Bare Ground Tundra	0.00
Wet Tundra	0.00
Mixed Tundra	0.00
Perennial Snowfields or Glaciers	0.00

SUBROUTINES CALLED: (none)

DESCRIPTION:

This is frequently used to detect cells at the boundaries of cell map. The tally of the distinct values includes cells with null (also referred to as NODATA) values. This additional inclusion over `param_focalvariety-data.aml` enables the detection of cells that are at the edge of the AOI.

REFERENCES: (none)

param_gwcell-col_id.aml

DEFINITION: Determines the column number associated within each zone of the input zone map. The input zone map is assumed to have been created by the **Tool Panel>>Automated Methods>>Squares (NNNX-NNNY)** tool.

ORIGINAL PURPOSE: (generic)

EXTRA DATA USED: (none)

SUBROUTINES CALLED: (none)

DESCRIPTION:

The **Tool Panel>>Automated Methods>>Squares (NNNX-NNNY)** tool will create a map of square zones. Each zone is uniquely identified. Also attached to each zone is a column number and a row number. These additional identifiers are inserted into the standard VAT associated with the output GRID and are referred to as NNNX and NNNY, respectively. The `param_gwcell-col_id.aml` extracts the NNNX information. If this information is not present in the VAT, then the parameterization method will fail.

The param_intersect-gwcell-col_id.aml method will intersect an input zone map that has not been created with the **Tool Panel>>Automated Methods>>Squares (NNNX-NNNY)** tool with one that has, and determine the column identifier of the square in that second GRID that most underlies each zone in the input zone map. Please see the description for the `param_intersect-gwcell-col_id.aml` method for more information.

This is a companion method to `param_gwcell-row_id.aml`.

REFERENCES: (none)

param_gwcell-row_id.aml

DEFINITION: Determines the row number associated within each zone of the input zone map. The input zone map is assumed to have been created by the **Tool Panel>>Automated Methods>>Squares (NNNX-NNNY)** tool.

ORIGINAL PURPOSE: (generic)

EXTRA DATA USED: (none)

SUBROUTINES CALLED: (none)

DESCRIPTION:

The **Tool Panel>>Automated Methods>>Squares (NNNX-NNNY)** tool will create a map of square zones. Each zone is uniquely identified. Also attached to each zone is a column number and a row number. These additional identifiers are inserted into the standard VAT associated with the output GRID and are referred to as NNNX and NNNY, respectively. The `param_gwcell-row_id.aml` extracts the NNNY information. If this information That the `param_intersect-gwcell-row_id.aml` method will intersect an input zone map that has not been created with the **Tool Panel>>Automated Methods>>Squares (NNNX-NNNY)** tool with one that has, and determine the column identifier of the square in that second GRID that most underlies each zone in the input zone map. Please see the description for the `param_intersect-gwcell-row_id.aml` method for more information.

This is a companion method to `param_gwcell-col_id.aml`.

REFERENCES: (none)

param_id.aml

DEFINITION: Reports the identification number of each zone in the input zone map.

ORIGINAL PURPOSE: `segment_order` (used for an unpublished Muskingum-routing model)

EXTRA DATA USED: (none)

SUBROUTINES CALLED: (none)

DESCRIPTION: (none)

REFERENCES: (none)

param_imperv.aml

DEFINITION: Determines the mean percentage of imperviousness for each zone in the GRID of polygon-type features, expressed in percent.

Table 6. Reclassification scheme for converting land use/land cover into inches of (winter) snow interception.

Input land use/land cover	Output interception of (winter) snow (inches)
White-red-jack pine	0.10
Spruce-fir	0.10
Longleaf-slash pine	0.10
Loblolly-shortleaf pine	0.10
Oak-pine	0.07
Oak-hickory	0.02
Oak-gum-cypress	0.02
Elm-ash-cottonwood	0.02
Maple-beech-birch	0.02
Aspen-birch	0.02
Douglas-fir	0.10
Hemlock-Sitka spruce	0.10
Ponderosa pine	0.10
Western white pine	0.10
Lodgepole pine	0.10
Larch	0.02
Fir-spruce	0.10
Redwood	0.10
Chaparral	0.05
Pinyon-juniper	0.10
Western hardwoods	0.02
Aspen-birch	0.02
Nonforest	0.00
Urban or Built-up Land	0.00
Dryland Cropland and Pasture	0.00
Irrigated Cropland and Pasture	0.00
Mixed Dryland/Irrigated Cropland and Pasture	0.00
Cropland/Grassland Mosaic	0.00
Cropland/Woodland Mosaic	0.03
Grassland	0.00
Shrubland	0.02
Mixed Shrubland/Grassland	0.02
Chaparral	0.02
Savanna	0.02
Broadleaf Deciduous Forest	0.02
Evergreen Coniferous Forest	0.10
Subalpine Forest	0.10
Mixed Forest	0.07
Deciduous Coniferous Forest	0.02
Evergreen Broadleaf Forest	0.03
Water Bodies	0.00
Herbaceous Wetland	0.02
Forested Wetland	0.10
Barren or Sparsely Vegetated	0.00
Wooded Tundra	0.02
Herbaceous Tundra	0.02
Bare Ground Tundra	0.00
Wet Tundra	0.00
Mixed Tundra	0.00
Perennial Snowfields or Glaciers	0.00

ORIGINAL PURPOSE: `hru_percent_imperv` (PRMS, although not currently in use)

EXTRA DATA USED: (none)

SUBROUTINES CALLED:

   `sfc_imperv.aml`

DESCRIPTION:

The subroutine, `sfc_imperv.aml`, will reclassify the land use/land cover GRID stored in the data bin (`%.par$data_bin%/lcov_comp/lulc`) into a per-cell estimate of percent of land cover that is impervious. The parameterization method (`param_imperv.aml`) will find the mean value of the cells within each zone in the input zone map.

This routine was originally designed for deriving `hru_percent_imperv` for the PRMS watershed model. It is not currently in use because the quality of GIS data that are available for deriving this information was not of sufficient quality.

The default reclassification of land cover into levels of impermeability is defined according to table 5. If this method is run noninteractively, then this default scheme is used.

REFERENCES:

Anderson, J.R., Hardy, E.E., Roach, J.T., and Witmer, R.E., 1976, A land use and land cover classification system for use with remote sensor data: U.S. Geological Survey Professional Paper 964, 28 p.
Leavesley, G.H., Lichty, R.W., Troutman, B.M., and Saindon, L.G., 1983, Precipitation-Runoff Modeling System - User's Manual: U.S. Geological Survey Water-Resources Investigations 83-4238, 207 p.

param_inflow-primary.aml

DEFINITION: Returns the identity of the largest zone that is immediately upstream from each zone in the input zone map.

ORIGINAL PURPOSE: `upst_inflow1` (PRMS storm mode)

EXTRA DATA USED: (none)

SUBROUTINES CALLED:

Table 7. Reclassification scheme for converting land use/land cover into inches of summer rain interception.

Input land use/land cover	Output interception of summer rain (inches)
White-red-jack pine	0.05
Spruce-fir	0.05
Longleaf-slash pine	0.05
Loblolly-shortleaf pine	0.05
Oak-pine	0.05
Oak-hickory	0.05
Oak-gum-cypress	0.05
Elm-ash-cottonwood	0.05
Maple-beech-birch	0.05
Aspen-birch	0.05
Douglas-fir	0.05
Hemlock-Sitka spruce	0.05
Ponderosa pine	0.05
Western white pine	0.05
Lodgepole pine	0.05
Larch	0.05
Fir-spruce	0.05
Redwood	0.05
Chaparral	0.05
Pinyon-juniper	0.05
Western hardwoods	0.05
Aspen-birch	0.05
Nonforest	0.00
Urban or Built-up Land	0.00
Dryland Cropland and Pasture	0.02
Irrigated Cropland and Pasture	0.02
Mixed Dryland/Irrigated Cropland and Pasture	0.02
Cropland/Grassland Mosaic	0.02
Cropland/Woodland Mosaic	0.05
Grassland	0.05
Shrubland	0.05
Mixed Shrubland/Grassland	0.05
Chaparral	0.05
Savanna	0.05
Broadleaf Deciduous Forest	0.05
Evergreen Coniferous Forest	0.05
Subalpine Forest	0.05
Mixed Forest	0.05
Deciduous Coniferous Forest	0.05
Evergreen Broadleaf Forest	0.05
Water Bodies	0.00
Herbaceous Wetland	0.05
Forested Wetland	0.05
Barren or Sparsely Vegetated	0.00
Wooded Tundra	0.05
Herbaceous Tundra	0.05
Bare Ground Tundra	0.00
Wet Tundra	0.02
Mixed Tundra	0.02
Perennial Snowfields or Glaciers	0.00

`inflow-ranking.aml`

DESCRIPTION:

   This routine is designed to be applied to a GRID of a drainage network. Each link in the network is examined to determine the identity of the links that flow into it. The link with the greatest flow accumulation is determined and reported by `param_inflow-primary.aml`.

REFERENCES:
Leavesley, G.H., Lichty, R.W., Troutman, B.M., and Saindon, L.G., 1983, Precipitation-Runoff Modeling System - User's Manual: U.S. Geological Survey Water-Resources Investigations 83-4238, 207 p.

param_inflow-secondary.aml

   DEFINITION: Returns the identity of the second largest zone that is immediately upstream to each zone in the input zone map.

   ORIGINAL PURPOSE: `upst_inflow2` (PRMS storm mode)

   EXTRA DATA USED: (none)

   SUBROUTINES CALLED:

      `inflow-ranking.aml`

DESCRIPTION:

   This routine is designed to be applied to a GRID of a drainage network. Each link in the network is examined to determine the identity of the links that flow into it. The link with the second greatest flow accumulation is determined and reported by this `param_inflow-secondary.aml`.

REFERENCES:
Leavesley, G.H., Lichty, R.W., Troutman, B.M., and Saindon, L.G., 1983, Precipitation-Runoff Modeling System - User's Manual: U.S. Geological Survey Water-Resources Investigations 83-4238, 207 p.

param_inflow-tertiary.aml

   DEFINITION: Returns the identity of the third largest zone that is immediately upstream from each zone in the input zone map.

   ORIGINAL PURPOSE: `upst_inflow3` (PRMS storm mode)

Table 8. Reclassification scheme for converting land use/land cover into inches of winter rain interception.

Input land use/land cover	Output interception of winter rain (inches)
White-red-jack pine	0.05
Spruce-fir	0.05
Longleaf-slash pine	0.05
Loblolly-shortleaf pine	0.05
Oak-pine	0.03
Oak-hickory	0.02
Oak-gum-cypress	0.02
Elm-ash-cottonwood	0.02
Maple-beech-birch	0.02
Aspen-birch	0.02
Douglas-fir	0.05
Hemlock-Sitka spruce	0.05
Ponderosa pine	0.05
Western white pine	0.05
Lodgepole pine	0.05
Larch	0.02
Fir-spruce	0.05
Redwood	0.05
Chaparral	0.02
Pinyon-juniper	0.05
Western hardwoods	0.02
Aspen-birch	0.02
Nonforest	0.00
Urban or Built-up Land	0.00
Dryland Cropland and Pasture	0.02
Irrigated Cropland and Pasture	0.02
Mixed Dryland/Irrigated Cropland and Pasture	0.02
Cropland/Grassland Mosaic	0.02
Cropland/Woodland Mosaic	0.05
Grassland	0.05
Shrubland	0.05
Mixed Shrubland/Grassland	0.05
Chaparral	0.05
Savanna	0.05
Broadleaf Deciduous Forest	0.02
Evergreen Coniferous Forest	0.05
Subalpine Forest	0.05
Mixed Forest	0.03
Deciduous Coniferous Forest	0.02
Evergreen Broadleaf Forest	0.03
Water Bodies	0.00
Herbaceous Wetland	0.05
Forested Wetland	0.05
Barren or Sparsely Vegetated	0.00
Wooded Tundra	0.05
Herbaceous Tundra	0.05
Bare Ground Tundra	0.00
Wet Tundra	0.02
Mixed Tundra	0.02
Perennial Snowfields or Glaciers	0.00

EXTRA DATA USED: (none)

SUBROUTINES CALLED:

    inflow-ranking.aml

DESCRIPTION:

This routine is designed to be applied to a GRID of a drainage network. Each link in the network is examined to determine the identity of the links that flow into it. The link with the third greatest flow accumulation is determined and reported by this param_inflow-secondary.aml.

REFERENCES:
Leavesley, G.H., Lichty, R.W., Troutman, B.M., and Saindon, L.G., 1983, Precipitation-Runoff Modeling System - User's Manual: U.S. Geological Survey Water-Resources Investigations 83-4238, 207 p.

param_intcp-mean-snow.aml

DEFINITION: Determines the average number of inches of interception of (winter-time) snow by all land covers for each zone in the input zone map.

ORIGINAL PURPOSE: snow_inctp (PRMS, replaced by param_intcp-snow2.aml)

EXTRA DATA USED: (none)

SUBROUTINES CALLED:

    sfc_intcp-snow.aml

DESCRIPTION:

This parameterization method will generate the average value (per-zone) of the input zone map) from the per-cell GRID produced by the sfc_intcp-snow.aml subroutine. The per-cell GRID is created by reclassifying the data bin-held land use/land cover GRID (%.par$data_bin%/lcov_comp/lulc) using the reclassification scheme shown in table 6.

REFERENCES:

Anderson, J.R., Hardy, E.E., Roach, J.T., and Witmer, R.E., 1976, A land use and land cover classification system for use with remote sensor data: U.S. Geological Survey Professional Paper 964, 28 p.

Leavesley, G.H., Lichty, R.W., Troutman, B.M., and Saindon, L.G., 1983, Precipitation-Runoff Modeling System - User's Manual: U.S. Geological Survey Water-Resources Investigations 83-4238, 207 p.

Powell, D.S., Faulkner, J.L., Darr, D.R., Zhu, Zhiliang, and MacCleery, D.W., 1993, Forest resources of the United States, 1992, General Technical Report RM-234 (revised): Fort Collins, Colorado, U.S. Department of Agriculture, Forest Service, 132 p.

param_intcp-mean-srain.aml

DEFINITION: Determines the average number of inches of interception of summer rain by all land cover for each zone in the input zone map.

ORIGINAL PURPOSE: srain_inctp (PRMS, replaced by param_intcp-srain2.aml)

EXTRA DATA USED: (none)

SUBROUTINES CALLED:

sfc_intcp-srain.aml

DESCRIPTION:

This parameterization method will generate the average value (per-zone) of the input zone map) from the per-cell GRID produced by the sfc_intcp-srain.aml subroutine. The per-cell GRID is created by reclassifying the data bin-held land use/land cover GRID (%.par$data_bin%/lcov_comp/lulc), using the reclassification scheme shown in table 7.

REFERENCES:

Anderson, J.R., Hardy, E.E., Roach, J.T., and Witmer, R.E., 1976, A land use and land cover classification system for use with remote sensor data: U.S. Geological Survey Professional Paper 964, 28 p.

Leavesley, G.H., Lichty, R.W., Troutman, B.M., and Saindon, L.G., 1983, Precipitation-Runoff Modeling System - User's Manual: U.S. Geological Survey Water-Resources Investigations 83-4238, 207 p.

Powell, D.S., Faulkner, J.L., Darr, D.R., Zhu, Zhiliang, and MacCleery, D.W., 1993, Forest resources of the United States, 1992, General Technical Report RM-234 (revised): Fort Collins, Colorado, U.S. Department of Agriculture, Forest Service, 132 p.

param_intcp-mean-wrain.aml

DEFINITION: Determines the average number of inches of interception of winter rain by all land covers for each zone in the input zone map.

ORIGINAL PURPOSE: wrain_inctp (PRMS, replaced by param_intcp-wrain2.aml)

EXTRA DATA USED: (none)

SUBROUTINES CALLED:

    sfc_intcp-srain.aml

DESCRIPTION:

This parameterization method will generate the average value (per-zone) of the input zone map) from the per-cell GRID produced by the `sfc_intcp-wrain.aml` subroutine. The per-cell GRID is created by reclassifying the data bin-held land use/land cover GRID (`%.par$data_bin%/lcov_comp/lulc`), using the reclassification scheme shown in table 8.

REFERENCES:

Anderson, J.R., Hardy, E.E., Roach, J.T., and Witmer, R.E., 1976, A land use and land cover classification system for use with remote sensor data: U.S. Geological Survey Professional Paper 964, 28 p.

Leavesley, G.H., Lichty, R.W., Troutman, B.M., and Saindon, L.G., 1983, Precipitation-Runoff Modeling System - User's Manual: U.S. Geological Survey Water-Resources Investigations 83-4238, 207 p.

Powell, D.S., Faulkner, J.L., Darr, D.R., Zhu, Zhiliang, and MacCleery, D.W., 1993, Forest resources of the United States, 1992, General Technical Report RM-234 (revised): Fort Collins, Colorado, U.S. Department of Agriculture, Forest Service, 132 p.

param_intcp-snow.aml

DEFINITION: Determines the number of inches of interception of (winter-time) snow by the *dominant* vegetation types for each zone in the input zone map.

ORIGINAL PURPOSE: `snow_inctp` (PRMS, replaced by `param_intcp-snow2.aml`)

EXTRA DATA USED:

    ArcInfo GRID of tree species from the data bin (`.../data_bin/forests/lower48`)

SUBROUTINES CALLED:

    param_majority.aml

DESCRIPTION:

This parameterization method will use the `param_majority.aml` parameterization method to determine the dominant tree type (nontree land covers are not included in this analysis!). The `param_intcp-snow.aml` method then reclassifies the dominant tree type, using the reclassification scheme shown in table 6 as the default, into a number of inches indicating the depth of snow interception.

REFERENCES:

Anderson, J.R., Hardy, E.E., Roach, J.T., and Witmer, R.E., 1976, A land use and land cover classification system for use with remote sensor data: U.S. Geological Survey Professional Paper 964, 28 p.
Leavesley, G.H., Lichty, R.W., Troutman, B.M., and Saindon, L.G., 1983, Precipitation-Runoff Modeling System - User's Manual: U.S. Geological Survey Water-Resources Investigations 83-4238, 207 p.
Powell, D.S., Faulkner, J.L., Darr, D.R., Zhu, Zhiliang, and MacCleery, D.W., 1993, Forest resources of the United States, 1992, General Technical Report RM-234 (revised): Fort Collins, Colorado, U.S. Department of Agriculture, Forest Service, 132 p.

param_intcp-snow2.aml

DEFINITION: Determines the average number of inches of interception of (winter-time) snow based on the *dominant PRMS cover type* for each zone in the input zone map.

ORIGINAL PURPOSE: `snow_inctp` (PRMS)

EXTRA DATA USED:

ArcInfo GRID of tree species from the data bin (.../`data_bin/forests/lower48`)

SUBROUTINES CALLED:

`sfc_cov-type-prms.aml`

`param_cov-type-prms3.aml`

DESCRIPTION:

This parameterization method will determine the PRMS cover type on a per-cell basis using the `sfc_cov-type-prms.aml` subroutine. The dominant PRMS cover type within each zone of the input zone map will be determined by `param_cov-type-prms3.aml`. Where the per-cell map identifies locations whose PRMS cover type are consistent with the per-zone dominant PRMS cover type, an interception is assigned.

Where the first four cover types in the PRMS cover type series, [bare, grass, shrub, tree], are found on a per-cell basis, interception values are assigned [`0.00`, `0.02`, `0.05`], respectively. For all other areas (that is, where the per-cell PRMS cover type matches "tree") the tree species map is reclassified according to the scheme shown in table 6. A per-zone average of these interception rates is derived for each zone in the input zone map.

REFERENCES:

Anderson, J.R., Hardy, E.E., Roach, J.T., and Witmer, R.E., 1976, A land use and land cover classification system for use with remote sensor data: U.S. Geological Survey Professional Paper 964, 28 p.

Leavesley, G.H., Lichty, R.W., Troutman, B.M., and Saindon, L.G., 1983, Precipitation-Runoff Modeling System - User's Manual: U.S. Geological Survey Water-Resources Investigations 83-4238, 207 p.

Powell, D.S., Faulkner, J.L., Darr, D.R., Zhu, Zhiliang, and MacCleery, D.W., 1993, Forest resources of the United States, 1992, General Technical Report RM-234 (revised): Fort Collins, Colorado, U.S. Department of Agriculture, Forest Service, 132 p.

param_intcp-srain.aml

DEFINITION: Determines the number of inches of interception of summer rain by the *dominant* vegetation types for each zone in the input zone map.

ORIGINAL PURPOSE: `srain_inctp` (PRMS, replaced by `param_intcp-srain2.aml`)

EXTRA DATA USED:

ArcInfo GRID of tree species from the data bin (`.../data_bin/forests/lower48`)

SUBROUTINES CALLED:

`param_majority.aml`

DESCRIPTION:

This parameterization method will use the `param_majority.aml` parameterization method to determine the dominant tree type (nontree land covers are not included in this analysis!). The `param_intcp-srain.aml` method then reclassifies the dominant tree type, using the reclassification scheme shown in table 7 as the default, into a number of inches indicating the depth of summer rain interception.

REFERENCES:

Anderson, J.R., Hardy, E.E., Roach, J.T., and Witmer, R.E., 1976, A land use and land cover classification system for use with remote sensor data: U.S. Geological Survey Professional Paper 964, 28 p.

Leavesley, G.H., Lichty, R.W., Troutman, B.M., and Saindon, L.G., 1983, Precipitation-Runoff Modeling System - User's Manual: U.S. Geological Survey Water-Resources Investigations 83-4238, 207 p.

Powell, D.S., Faulkner, J.L., Darr, D.R., Zhu, Zhiliang, and MacCleery, D.W., 1993, Forest resources of the United States, 1992, General Technical Report RM-234 (revised): Fort Collins, Colorado, U.S. Department of Agriculture, Forest Service, 132 p.

param_intcp-srain2.aml

DEFINITION: Determines the average number of inches of interception of summer rain based on the per-cell *PRMS cover type* in a zone, for each zone in the input zone map.

ORIGINAL PURPOSE: `srain_inctp` (PRMS)

EXTRA DATA USED:

ArcInfo GRID of tree species (for example, `%.par$data_bin%/forests/lower48`)

SUBROUTINES CALLED:

`sfc_cov-type-prms.aml`

`param_cov-type-prms3.aml`

DESCRIPTION:

This parameterization method will determine the PRMS cover type on a per-cell basis using the `sfc_cov-type-prms.aml` subroutine. The dominant PRMS cover type within each zone of the input zone map will be determined by `param_cov-type-prms3.aml`.

Where the first four cover types in the PRMS cover type series, [bare, grass, shrub, tree], are found on a per-cell basis, interception values are assigned [0.00, 0.02, 0.05], respectively. Where the per-cell PRMS cover type matches "tree," the tree species map is reclassified according to the scheme shown in table 7. A per-zone average of these interception rates is derived for each zone in the input zone map.

REFERENCES:

Anderson, J.R., Hardy, E.E., Roach, J.T., and Witmer, R.E., 1976, A land use and land cover classification system for use with remote sensor data: U.S. Geological Survey Professional Paper 964, 28 p.
Leavesley, G.H., Lichty, R.W., Troutman, B.M., and Saindon, L.G., 1983, Precipitation-Runoff Modeling System - User's Manual: U.S. Geological Survey Water-Resources Investigations 83-4238, 207 p.
Powell, D.S., Faulkner, J.L., Darr, D.R., Zhu, Zhiliang, and MacCleery, D.W., 1993, Forest resources of the United States, 1992, General Technical Report RM-234 (revised): Fort Collins, Colorado, U.S. Department of Agriculture, Forest Service, 132 p.

param_intcp-srain2-meters.aml

DEFINITION: Produces the meters-equivalent of the output produced by `param_intcp-srain2.aml`.

ORIGINAL PURPOSE: (generic)

EXTRA DATA USED: (none)

SUBROUTINES CALLED:

    param_intcp-srain2.aml

DESCRIPTION:

See the description for `param_intcp-srain2.aml`. `Param_intcp-srain2-meters.aml` takes the output of that method and converts it to meters.

REFERENCES: (none)

param_intcp-wrain.aml

DEFINITION: Determines the number of inches of interception of winter rain by the *dominant* tree species for each zone in the input zone map.

ORIGINAL PURPOSE: `wrain_inctp` (PRMS, replaced by `param_intcp-wrain2.aml`)

EXTRA DATA USED:

    ArcInfo GRID of tree species from the data bin (.../data_bin/forests/lower48)

SUBROUTINES CALLED:

    param_majority.aml

DESCRIPTION:

This parameterization method will use the `param_majority.aml` parameterization method to determine the dominant tree type (nontree land covers are not included in this analysis!). The `param_intcp-wrain.aml` method then reclassifies the dominant tree type, using the reclassification scheme shown in table 8 as the default, into a number of inches indicating the depth of winter rain interception.

REFERENCES:

Anderson, J.R., Hardy, E.E., Roach, J.T., and Witmer, R.E., 1976, A land use and land cover classification system for use with remote sensor data: U.S. Geological Survey Professional Paper 964, 28 p.
Leavesley, G.H., Lichty, R.W., Troutman, B.M., and Saindon, L.G., 1983, Precipitation-Runoff Modeling System - User's Manual: U.S. Geological Survey Water-Resources Investigations 83-4238, 207 p.

Powell, D.S., Faulkner, J.L., Darr, D.R., Zhu, Zhiliang, and MacCleery, D.W., 1993, Forest resources of the
United States, 1992, General Technical Report RM-234 (revised): Fort Collins, Colorado, U.S. Department
of Agriculture, Forest Service, 132 p.

param_intcp-wrain2.aml

DEFINITION: Determines the average number of inches of interception of winter rain based on the per-
cell *PRMS cover type* in a zone, for each zone in the input zone map.

ORIGINAL PURPOSE: `wrain_inctp` (PRMS)

EXTRA DATA USED:

ArcInfo GRID of tree species from the data bin (`.../data_bin/forests/lower48`)

SUBROUTINES CALLED:

`sfc_cov-type-prms.aml`

`param_cov-type-prms3.aml`

DESCRIPTION:

This parameterization method will determine the PRMS cover type on a per-cell basis using the `sfc_
cov-type-prms.aml` subroutine. The dominant PRMS cover type within each zone of the input zone map
will be determined by `param_cov-type-prms3.aml`.

Where the first four cover types in the PRMS cover type series, [bare, grass, shrub, tree], are found on a per-
cell basis, interception values are assigned [0.00, 0.02, 0.05], respectively. Where the per-cell PRMS
cover type matches "tree," the tree species map is reclassified according to the scheme shown in table 8. A
per-zone average of these interception rates is derived for each zone in the input zone map.

REFERENCES:

Anderson, J.R., Hardy, E.E., Roach, J.T., and Witmer, R.E., 1976, A land use and land cover classification sys-
tem for use with remote sensor data: U.S. Geological Survey Professional Paper 964, 28 p.
Leavesley, G.H., Lichty, R.W., Troutman, B.M., and Saindon, L.G., 1983, Precipitation-Runoff Modeling System
- User's Manual: U.S. Geological Survey Water-Resources Investigations 83-4238, 207 p.
Powell, D.S., Faulkner, J.L., Darr, D.R., Zhu, Zhiliang, and MacCleery, D.W., 1993, Forest resources of the
United States, 1992, General Technical Report RM-234 (revised): Fort Collins, Colorado, U.S. Department
of Agriculture, Forest Service, 132 p.

param_intcp-wrain2-meters.aml

DEFINITION: Produces the meters equivalent of the output produced by `param_intcp-wrain2.aml`.

ORIGINAL PURPOSE: (generic)

EXTRA DATA USED: (none)

SUBROUTINES CALLED:

   `param_intcp-srain2.aml`

DESCRIPTION:

See the description for `param_intcp-wrain2.aml`. `param_intcp-wrain2-meters.aml` takes the output of that method and converts it to meters.

REFERENCES: (none)

param_intersect-gwcell-col_id.aml

DEFINITION: Intersects the input zone map with a second GRID that is assumed to have been created by the **Tool Panel>>Automated Methods>>Squares (NNNX-NNNY)** tool. Determines the zone in the second GRID that shares the largest area with each zone in the input zone map, and reports the column number associated with that zone.

ORIGINAL PURPOSE: (generic)

EXTRA DATA USED:

   ArcInfo GRID, user-specified output from **Tool Panel>>Automated Methods>>Squares (NNNX-NNNY) tool**.

SUBROUTINES CALLED: (none)

DESCRIPTION:

The **Tool Panel>>Automated Methods>>Squares (NNNX-NNNY)** tool will create a map of square zones. Each zone is uniquely identified. Also attached to each zone is a column number and a row number. These additional identifiers are inserted into the standard VAT associated with the output GRID and are referred to as NNNX and NNNY, respectively. The `param_intersect-gwcell-col_id.aml` extracts the NNNX information. If this information is not present in the VAT, then the parameterization method will fail.

The `param_gwcell-col_id.aml` method will not attempt to intersect an input zone map with another GRID. It will attempt to determine the column identifier of the zones in the input zone map (that is, assumes that the input zone map was created with the **Tool Panel>>Automated Methods>>Squares (NNNX-NNNY)** tool). Please see the description for the `param_gwcell-col_id.aml` method for more information.

REFERENCES: (none)

param_intersect-gwcell-row_id.aml

DEFINITION: Intersects the input zone map with a second GRID that is assumed to have been created by the **Tool Panel>>Automated Methods>>Squares (NNNX-NNNY)** tool. Determines the zone in the second GRID that shares the largest number of cells with each zone in the input zone map, and reports the row number associated that zone.

ORIGINAL PURPOSE: (generic)

EXTRA DATA USED:

ArcInfo GRID, user-specified output from **Tool Panel>>Automated Methods>>Squares (NNNX-NNNY)** tool.

SUBROUTINES CALLED: (none)

DESCRIPTION:

The **Tool Panel>>Automated Methods>>Squares (NNNX-NNNY)** tool will create a map of square zones. Each zone is uniquely identified. Also attached to each zone is a column number and a row number. These additional identifiers are inserted into the standard VAT associated with the output GRID and are referred to as NNNX and NNNY, respectively. The `param_intersect-gwcell-row_id.aml` extracts the NNNY information. If this information is not present in the VAT, then the parameterization method will fail.

The `param_gwcell-row_id.aml` method will not attempt to intersect an input zone map with another GRID. It will attempt to determine the column identifier of the zones in the input zone map (that is, assumes that the input zone map was created with the **Tool Panel>>Automated Methods>>Squares (NNNX-NNNY)** tool). Please see the description for the `param_gwcell-row_id.aml` method for more information.

REFERENCES: (none)

param_jh-coef.aml

DEFINITION: Estimate the mean Jenson-Haise coefficient for each zone in the input zone map, expressed in percent (0.00 - 1.00).

ORIGINAL PURPOSE: `jh_coef_hru` (PRMS)

EXTRA DATA USED:

SUBROUTINES CALLED:

`sfc_jh-coef.aml`

DESCRIPTION:

This method will determine the Jenson-Haise coefficient for each zone in the input zone map. The Jenson-Haise coefficient is derived according to

$$coef_{JH} = (27.5 - 0.25*(jh_2 - jh_1) - (jh_{elv}/1000)), \tag{1}$$

where jh_1 and jh_2 are the minimum and maximum saturation vapor pressures, respectively. jh_{elv} is defined as the median elevation, expressed in feet, for the area for which the Jenson-Haise coefficient is being derived. The median elevation for each zone automatically calculated by the method during determination of the Jenson-Haise coefficient.

Saturation vapor pressure is calculated by

$$jh_i = 10*0.61078*e^{(17.269*T_i)/T_i+237.30)}, \tag{2}$$

where jh_i refers to either jh_1 or jh_2. When calculating jh_1, T_i is the mean minimum temperature (degrees Celsius) for the warmest month of the year. When calculating jh_2, T_i is the mean maximum temperature (degrees Celsius) for the warmest month of the year. The default mean minimum temperature for the warmest month is 7 degrees Centigrade. The default mean maximum temperature for the warmest month is 25 degrees Celsius. The user may modify these values through a worksheet menu if the parameterization method is run interactively. From this worksheet, the user may change the mean minimum and maximum temperatures for the warmest month. The work sheet will automatically display the Jenson-Haise coefficient value for these temperatures and an arbitrary elevation value, which also may be modified. It should be noted that although the user may modify the elevation used in the worksheet calculation of a sample Jenson-Haise value, the elevation values used in the actual parameter derivation are derived as the median elevation for each zone in the input zone map. The user has no control, other than to actually modify the GRID of elevation, over the median elevation values used in the derivation of the Jenson-Haise equation.

REFERENCES:

Anderson, J.R., Hardy, E.E., Roach, J.T., and Witmer, R.E., 1976, A land use and land cover classification system for use with remote sensor data: U.S. Geological Survey Professional Paper 964, 28 p.

Leavesley, G.H., Lichty, R.W., Troutman, B.M., and Saindon, L.G., 1983, Precipitation-Runoff Modeling System - User's Manual: U.S. Geological Survey Water-Resources Investigations 83-4238, 207 p.

Powell, D.S., Faulkner, J.L., Darr, D.R., Zhu, Zhiliang, and MacCleery, D.W., 1993, Forest resources of the United States, 1992, General Technical Report RM-234 (revised): Fort Collins, Colorado, U.S. Department of Agriculture, Forest Service, 132 p.

param_leaf-loss.aml

DEFINITION: Estimates the average quantity of the reduction of the summer vegetation density with winter leaf loss, based on cover type, within each zone of the input zone map. Expressed in percent.

ORIGINAL PURPOSE: (generic)

EXTRA DATA USED:

SUBROUTINES CALLED:

```
sfc_leaf-loss.aml
```

DESCRIPTION:

The `sfc_leaf-loss.aml` subroutine will ensure that the data bin GRID for land use/land cover (found at .../data_bin/lcov_comp/lulc) is reclassified on a per-cell basis into the cover type scheme shown in table 1. The cover type GRID is then reclassified into a leaf loss factor using the scheme shown in table 3. The `param_leaf-loss.aml` method will then find the average of the per-cell leaf loss factors within each zone of the input zone map.

REFERENCES: (none)

param_line-slope.aml

DEFINITION: Determines the mean slope of each zone in the input zone map, expressed in percent.

ORIGINAL PURPOSE: (generic)

EXTRA DATA USED:

ArcInfo GRID of elevation, referred to as `%.aGrid%`

SUBROUTINES CALLED:

```
zone_flowlength.aml
```

```
param_range.aml
```

DESCRIPTION:

This method is intended to be used with an input zone map that depicts a drainage network. For each zone, the total range in elevation will be divided by the range in the flow length.

REFERENCES: (none)

param_loni-nbins.aml

DEFINITION: Set the number of categories ("bins") used to reclassify the (per-cell) values of the TOPMODEL topographic wetness index for each zone in the input zone map.

ORIGINAL PURPOSE: `nacsc` (TOPMODEL, replaced by `param_loni-nbins3.aml`)

EXTRA DATA USED: (none)

SUBROUTINES CALLED:

`sfc_loni.aml`

DESCRIPTION:

TOPMODEL requires a per-sub catchment description of the distribution of the topographic wetness index. The topographic wetness index (referred to here as the *loni*) is calculated on a per-cell basis using

$$loni = \ln((fac + 1)/slp), \tag{3}$$

where *fac* is the flow accumulation, or the number of upslope cells, and *slp* is the slope of a plane fitted to the eight neighbors of a cell, expressed in degrees. *slp* has an minimum value of 0.005 degrees to prevent division by zero. The derivation of *loni*. *fac* may be derived one of three ways: (1) accumulating on a cell by cell basis across the entire catchment, (2) accumulating on a cell by cell basis across the entire catchment, except for along the primary drainage path, and (3) accumulating on a cell by cell basis within each subcatchment. The last method attempts to eliminate the effect of upstream subcatchments on the distribution of loni for a downstream subcatchment. The first option is standard.

Rather than reporting all loni values to TOPMODEL, the values in the loni surface are reclassified or binned into ranges of values, and the proportion of the subcatchment area whose loni value is within each range is reported. The range of values that each bin represents varies according to the actual range of loni values found within the entire subcatchment and the number of bins that are used to describe the loni distribution for that subcatchment. Routine param_loni-nbins.aml, one of the first steps in this process, is used to set the number of bins for each zone (that is, subcatchment) in the input zone map.

The default number of loni bins is 30 categories. If this method is run in the interactive mode, the user is presented with a display of statistics about the distribution of loni for each zone in the input zone map and a second table which allows the user to reset the number of loni bins for individual subcatchments.

This method does not actually try to reclassify the loni surface. It merely sets the number of bins that will be used to reclassify the loni surface within each subcatchment. A subcatchment is defined here as

an area within which all runoff remains owing to topographic boundaries, such that this water is all ultimately channeled to one outlet.

(adapted from *www.hamiltonnature.org/habitats/glossary.htm*). Another way to define a subcatchment is as the single contributing area above some outlet point.

REFERENCES:

Beven, K. J., Quinn, P.F., Romanowicz, R.J., Freer, James, Fisher, J.I., and Lamb, R., 1995, TOPMODEL and GRIDATB – A Users Guide to the Distribution Versions (95.02): Lancaster University, United Kingdom, Centre for Research on Environmental Systems and Statistics, Institute of Environmental and Biological Sciences, 31 p.

param_loni-nbins2.aml

DEFINITION: Set the number of categories ("bins") used to reclassify the (per-cell) values of the TOPMODEL topographic wetness index for each zone in the input zone map.

ORIGINAL PURPOSE: `nacsc` (TOPMODEL, replaced by `param_loni-nbins3.aml`)

EXTRA DATA USED: (none)

SUBROUTINES CALLED:

```
zone_loni.aml
```

DESCRIPTION:

Please see the description for `param_loni-nbins.aml`. The *param_loni-nbins2.aml* method is an extension of that method. The important difference between this method and the former is that `param_loni-nbins2.aml` forces the usage of a localized flow accumulation in the derivation of the loni surface (that is, *fac* in equation 3), using the `zone_loni.aml` routine. This algorithm most closely approximates the third alternative for calculating flow accumulation in the description of `param_loni-nbins.aml`.

This method is designed to work against an input zone map that depicts subcatchments.

REFERENCES:

Beven, K. J., Quinn, P.F., Romanowicz, R.J., Freer, James, Fisher, J.I., and Lamb, R., 1995, TOPMODEL and GRIDATB – A Users Guide to the Distribution Versions (95.02): Lancaster University, United Kingdom, Centre for Research on Environmental Systems and Statistics, Institute of Environmental and Biological Sciences, 31 p.

param_loni-nbins3.aml

DEFINITION: Set the number of categories used reclassify the values of the TOPMODEL topographic index ("loni") for each zone in the GRID of polygon-type features.

ORIGINAL PURPOSE: `nacsc` (TOPMODEL)

EXTRA DATA USED: (none)

SUBROUTINES CALLED:


```
zone_loni.aml
```


```
zone_one-plane.aml
```


DESCRIPTION:

Please see the description for `param_loni-nbins2.aml`. The `param_loni-nbins3.aml` method is an extension of that method. The difference between this method and the former is that `param_loni-nbins3.aml` does not assume that the input zone map represents subcatchments. This method was designed to support more modern applications of TOPMODEL where subcatchments are potentially defined as an area between one side of a link in a drainage network (that is, one bank of the stream segment) and the topographic divide above that side of the link.

REFERENCES:

Beven, K. J., Quinn, P.F., Romanowicz, R.J., Freer, James, Fisher, J.I., and Lamb, R., 1995, TOPMODEL and GRIDATB – A Users Guide to the Distribution Versions (95.02): Lancaster University, United Kingdom, Centre for Research on Environmental Systems and Statistics, Institute of Environmental and Biological Sciences, 31 p.

param_loni-bin-ac.aml

DEFINITION: Returns the percent of enclosing subcatchment area that is covered by each category of a reclassified TOPMODEL topographic index ("loni") GRID, for each zone in the input zone map.

ORIGINAL PURPOSE: `ac` (TOPMODEL, not currently in use)

EXTRA DATA USED:

ArcInfo GRID of subcatchments (referred to as *nmru*)

SUBROUTINES CALLED: (none)

DESCRIPTION:

The input zone map to this method is assumed to have been produced by the **Tool Panel>>Automated Methods>>TOPMODEL-type Zones>>Loni Bins** tool. During the execution of the `param_loni-bin-ac.aml` method, the user will be prompted to specify the identity of the map of subcatchments that the map of loni bins (that is, the input zone map) was derived from.

The `param_loni-bin-ac.aml` method will copy an INFO table that was specially generated by this tool in order to generate parameter values. If this INFO table (which has the same name as the one given by the user for the tool output plus a ."loni" suffix) does not exist, then the parameterization method will produce no output.

REFERENCES:

Beven, K. J., Quinn, P.F., Romanowicz, R.J., Freer, James, Fisher, J.I., and Lamb, R., 1995, TOPMODEL and GRIDATB – A Users Guide to the Distribution Versions (95.02): Lancaster University, United Kingdom, Centre for Research on Environmental Systems and Statistics, Institute of Environmental and Biological Sciences, 31 p.

param_loni-bin-st-ac.aml

DEFINITION: Returns the percent of enclosing subcatchment area that is covered by each category of a reclassified TOPMODEL topographic index ("loni") GRID, for each zone in the input zone map. Also returns the minimum loni value associated with each loni category.

ORIGINAL PURPOSE: `st, ac` (TOPMODEL)

EXTRA DATA USED: (none)

SUBROUTINES CALLED: (none)

DESCRIPTION:

The input zone map to this method is assumed to have been produced by the **Tool Panel>>Automated Methods>>TOPMODEL-type Zones>>Loni Bins** tool. The `param_loni-bin-st-ac.aml` method will copy an

INFO table that was specially generated by this tool in order to generate parameter values. If this INFO table (which has the same name as the one given by the user for the tool output plus a `.loni` suffix) does not exist, then the parameterization method will produce no output.

This parameterization method is atypical when compared with other parameterization methods because it does not rely on the GIS Weasel Parameterization Engine to create the output. The parameterization method is responsible for creating an ASCII file output. This output file is separate from any other output from the GIS Weasel Parameterization Engine. The file has five columns. The first is the unique identifier associated with a particular loni bin within a particular subcatchment. This identifier is really used only by the GIS Weasel. The second column is the sequential identifier of the loni bin. These values are integers from 1 to the number of bins designated for the enclosing subcatchment. This column of information is also only used by the GIS Weasel. The third column is the subcatchment identifier. The fourth column is the "`st`" parameter that is the minimum value associated with a loni bin. The fifth column is the "`ac`" parameter that is the percentage of the subcatchment covered by the range of loni values represented by the corresponding `st` value.

Please see the descriptions for `param_loni-nbins.aml` and param_loni-nbins3.aml for more information.

REFERENCES:

Beven, K. J., Quinn, P.F., Romanowicz, R.J., Freer, James, Fisher, J.I., and Lamb, R., 1995, TOPMODEL and GRIDATB – A Users Guide to the Distribution Versions (95.02): Lancaster University, United Kingdom, Centre for Research on Environmental Systems and Statistics, Institute of Environmental and Biological Sciences, 31 p.

param_loni-mean.aml

DEFINITION: Returns the average loni value per zone of the input zone map.

ORIGINAL PURPOSE: (generic)

EXTRA DATA USED: (none)

SUBROUTINES CALLED:

    sfc_loni.aml

DESCRIPTION:

Please see the description of `param_loni-nbins.aml` for a discussion of the loni index. In particular, see equation 3.

REFERENCES: (none)

param_nac.aml

DEFINITION: Returns the identity of the most commonly occurring reclassified value of the loni surface within each zone of the input zone map.

ORIGINAL PURPOSE: (generic)

EXTRA DATA USED:

ArcInfo GRID of the loni bin identifiers

SUBROUTINES CALLED:

sfc_loni.aml

DESCRIPTION:

Please see the description of param_loni-nbins.aml for a discussion of the loni index. In particular, see equation 3. The user is asked to specify the ArcInfo GRID of the reclassified loni values during the execution of the method.

REFERENCES: (none)

param_n*.aml

DEFINITION: Returns the identity of the most commonly occurring value from a user specified ArcInfo GRID within each zone of the input zone map.

ORIGINAL PURPOSE: (generic, various)

EXTRA DATA USED:

ArcInfo GRID of identifiers

SUBROUTINES CALLED: (none)

DESCRIPTION:

This description applies to a group of parameterization methods that all function identically. Each will prompt the user during execution to specify the name of a secondary ArcInfo GRID that depicts and identifies some kind of geographic feature. Each parameterization method listed below is hard-coded to ask for a feature of a specific name. The parameterization method will then determine which zone in the secondary GRID underlies each zone in the input zone map the most, based on the number of cells that the two share. The parameterization methods include:

```
param_nchan.aml


param_ndabranch.aml


param_ndajunction.aml


param_ndanode.aml


param_ndanode-local.aml


param_nflowplane.aml


param_ngw.aml


param_ngwrow.aml


param_ngwcol.aml


param_nhru.aml


param_nlink.aml


param_nmru.aml


param_nradpl.aml


param_nreach.aml


param_ns.aml


param_nsc.aml
```

```
param_nshed.aml

param_nssr.aml

param_ntopchan.aml
```

While this set of methods is largely redundant, it does allow the developers to customize the prompts given to the user, greatly improving the ease of use of the software.

REFERENCES: (none)

param_nnnx-nnny-id.aml

DEFINITION: Returns the zone identifier, the column (that is, NNNX) value, and the row (that is, NNNY) value for each zone in the input zone map in an ASCII file. This parameterization method assumes the input zone map was created with the **Tool Panel>>Automated Methods>>Squares (NNNX-NNNY)** tool.

ORIGINAL PURPOSE: (generic)

EXTRA DATA USED: (none)

SUBROUTINES CALLED: (none)

DESCRIPTION:

This is a specialty parameterization method. It will produce ASCII file output without the aid of the Parameterization Engine (although it can run within the Parameterization Engine). It will produce a three-columned file. The first is the zone identifier, the second is the column number of the zone, and the third is the row number of the zone.

If the input zone map was not created with the **Tool Panel>>Automated Methods>>Squares (NNNX-NNNY)** tool, then the parameterization method will fail. See the description for the **Tool Panel>>Automated Methods>>Squares (NNNX-NNNY)** tool, in the "Delineation Phase" section, for more information.

REFERENCES: (none)

param_nnnx-nnny-nssr.aml

DEFINITION: Intersects the input zone map with a second GRID. The input zone map is assumed to have been created with the **Tool Panel>>Automated Methods>>Squares (NNNX-NNNY)** tool. For each intersection, the method returns the zone identifier, the column (that is, NNNX) value, and the row (that is, NNNY) value for the zone in the input zone map, and the identifier for each zone in a second GRID, in an ASCII file.

ORIGINAL PURPOSE: (generic)

EXTRA DATA USED: (none)

SUBROUTINES CALLED: (none)

DESCRIPTION:

This is a specialty parameterization method. It will produce ASCII file output without the aid of the Parameterization Engine (although it can run within the Parameterization Engine). It will produce a four-columned file. The first is the zone identifier, the second is the column number of the zone, the third is the row number of the zone, and the fourth is the identifier of the zones in the second GRID.

If the input zone map was not created with the **Tool Panel>>Automated Methods>>Squares (NNNX-NNNY)** tool, then the parameterization method will fail. See the description for the **Tool Panel>>Automated Methods>>Squares (NNNX-NNNY)** tool, in the "Delineation Phase" section, for more information.

REFERENCES: (none)

param_num-chan.aml

DEFINITION: Sets the number of channel-routing increments to be derived on a per-zone basis of the input zone map.

ORIGINAL PURPOSE: (generic, used in the delineation of TOPMODEL channel routine increments)

EXTRA DATA USED: (none)

SUBROUTINES CALLED:

```
zone_ntopchan-mainlink.aml
```

```
sfc_flowlength-down.aml
```

```
param_range.aml
```

DESCRIPTION:

The input zone map is assumed to be a map of TOPMODEL subcatchments. The `param_num-chan.aml` parameterization method will, by default, set the number of channel-routing increments to 5 if there are at least 10 cells in the main path of drainage in each zone as determined using `zone_ntopchan-mainlink.aml`. If this is not the case, then the number of cells in the main link of drainage is divided by 10 and integerized. A minimum value of 2 is allowed (although there is no check or warning to the user is the number of cells in the main link of drainage is less than 20 cells).

If this parameterization method is run in the interactive mode, then the user is presented with some supporting information and can edit the values prior to accepting them.

This parameterization method only sets the number of channel-routing increments that will be delineated. It does not delineate the channel-routing increments. This is done using **Tool Panel>>Automated Methods>>Topmodel-type Zones>>Channel Increments**.

REFERENCES: (none)

param_num-ndanode.aml

DEFINITION: Determines the number of zones in a second GRID, representing DAFLOW nodes, found in each of the zones of the input zone map.

ORIGINAL PURPOSE: `daf_num_xsec` (DAFLOW)

EXTRA DATA USED: (none)

SUBROUTINES CALLED: (none)

DESCRIPTION:

During the execution of this parameterization method, the user must specify the name of an ArcInfo GRID representing the DAFLOW nodes for the modeling effort. The number of nodes within each zone of the input zone map is determined. For more information on the DAFLOW model, please see Jobson (1989).

REFERENCES:

Jobson, H.E., 1989, Users manual for an open-channel streamflow model based on the diffusion analogy: U.S. Geological Survey Water-Resources Investigations Report 89-4133, 73 p.

param_ofpl-inflow-primary.aml

DEFINITION: Reports which of the zones in a second, user-specified, GRID occupies most of the contributing area for each zone of the input zone map. Expects the input zone map is a drainage network.

ORIGINAL PURPOSE: `lat_inflowl` (PRMS storm mode, replaced by `param_ofpl-inflow-primary2.aml`)

EXTRA DATA USED:

ArcInfo GRID of overland flow planes

SUBROUTINES CALLED:

`zone_one-plane.aml`

DESCRIPTION:

This routine is designed to be applied to a GRID of a drainage network. The contributing area to each link in the network is created using the `zone_one-plane.aml` routine. These contributing areas are intersected with a user-specified ArcInfo GRID, representing overland flow planes (see the PRMS documentation for more details on the overland flow plane concept). The identity of the overland flow plane in the largest intersection associated with each zone in the input zone map is reported.

The GRID produced by this parameterization method has the spatial extent of the contributing areas to the links in the drainage network (that is, the zones in the input zone map).

REFERENCES:

Leavesley, G.H., Lichty, R.W., Troutman, B.M., and Saindon, L.G., 1983, Precipitation-Runoff Modeling System User's Manual: U.S. Geological Survey Water-Resources Investigations Report 83-4238, p. 207.

param_ofpl-inflow-primary2.aml

DEFINITION: Reports which of the zones in a second, user-specified, GRID occupies the most of the contributing area derived automatically for each zone of the input zone map. Expects the input zone map is a drainage network.

ORIGINAL PURPOSE: `lat_inflowl` (PRMS storm mode)

EXTRA DATA USED:

ArcInfo GRID of overland flow planes

SUBROUTINES CALLED:

```
zone_one-plane.aml
```

DESCRIPTION:

See the description for `param_ofpl-inflow-primary.aml`. `param_ofpl-inflow-primary2.aml` is an evolution of this method. The difference is that rather than determining the dominant overland flow plane for the contributing area for each link, the dominant contributing area to each link per each overland flow plane is determined.

REFERENCES:

Leavesley, G.H., Lichty, R.W., Troutman, B.M., and Saindon, L.G., 1983, Precipitation-Runoff Modeling System User's Manual: U.S. Geological Survey Water-Resources Investigations Report 83-4238, p. 207.

param_ofpl-inflow-secondary.aml

DEFINITION: Reports which of the zones in a second, user-specified, GRID has the second-largest intersection with the contributing area derived automatically for each zone of the input zone map. Expects the input zone map is a drainage network.

ORIGINAL PURPOSE: `lat_inflowr` (PRMS storm mode, replaced by `param_ofpl-inflow-secondary2.aml`)

EXTRA DATA USED: (none)

SUBROUTINES CALLED:

```
zone_one-plane.aml
```

```
param_ofpl-inflow-primary.aml
```

DESCRIPTION:

This routine is designed to be applied to a GRID of a drainage network. The contributing area to each link in the network is created using the `zone_one-plane.aml` routine. These contributing areas are intersected with a user-specified ArcInfo GRID, representing overland flow planes (see the PRMS documentation (Leavesley and others, 1983) for more details on the overland flow plane concept).

The identity of the overland flow plane in the largest intersection associated with each zone in the input zone map is determined using `param_ofpl-inflow-primary.aml`. The `param_ofpl-inflow-secondary.aml` method notes the identity of this zone and excludes from it from the subsequent analysis of the largest intersection between the contributing area map and the overland flow plane map.

The GRID produced by this parameterization method has the spatial extent of the contributing areas to the links in the drainage network (that is, the zones in the input zone map).

REFERENCES:

Leavesley, G.H., Lichty, R.W., Troutman, B.M., and Saindon, L.G., 1983, Precipitation-Runoff Modeling System User's Manual: U.S. Geological Survey Water-Resources Investigations Report 83-4238, p. 207.

param_ofpl-inflow-secondary2.aml

DEFINITION: Reports which of the zones in a second, user-specified, GRID has the second-largest inter-section with the contributing area derived automatically for each zone of the input zone map. Expects the input zone map is a drainage network.

ORIGINAL PURPOSE: `lat_inflowr` (PRMS storm mode)

EXTRA DATA USED:

ArcInfo GRID of overland flow planes

SUBROUTINES CALLED:

`zone_one-plane.aml`

`param_ofpl-inflow-primary2.aml`

DESCRIPTION:

See the description for `param_ofpl-inflow-primary2.aml`; `param_ofpl-inflow-secondary2.aml` builds on this method. Using `param_ofpl-inflow-primary2.aml`, `param_ofpl-inflow-secondary2.aml` determines the dominant contributing area to each overland flow plane. Then, as was in the case of `param_ofpl-inflow-secondary.aml`, this contributing area is noted and eliminated from the subsequent analysis. Then `param_ofpl-inflow-secondary2.aml` determines the next most dominant contributing area per flow plane.

REFERENCES:

Leavesley, G.H., Lichty, R.W., Troutman, B.M., and Saindon, L.G., 1983, Precipitation-Runoff Modeling System User's Manual: U.S. Geological Survey Water-Resources Investigations Report 83-4238, p. 207.

param_ofpl-length.aml

DEFINITION: Reports the "overland flow plane length" of each zone in the input zone map. Expressed in the XY units of the input zone map.

ORIGINAL PURPOSE: `ofp_length` (PRMS storm mode)

EXTRA DATA USED:

ArcInfo GRID of overland flow planes

SUBROUTINES CALLED:

`param_flowlength.aml`

`zone_one-plane.aml`

`param_area.aml`

DESCRIPTION:

The input zone map is expected to represent overland flow planes. The user is prompted during the execution of the parameterization method to specify the name of the ArcInfo GRID of stream channels from which the overland flow plane map was generated. Each zone in the input zone map is associated with a zone in the stream channel map. The length of the associated stream channel is divided into the area of the zone to determine what is termed here as the *overland flow plane length*.

Overland flow planes are idealized in the computations of the PRMS storm-event model as rectangularly shaped features. The edge of the rectangle, along the associated stream channel, is assumed to be straight. The stream channel is assumed to be exactly as long as that side of the overland flow plane. The width, based on these assumptions, can be inferred based on the channel length and the area of the overland flow plane.

REFERENCES:

Leavesley, G.H., Lichty, R.W., Troutman, B.M., and Saindon, L.G., 1983, Precipitation-Runoff Modeling System User's Manual: U.S. Geological Survey Water-Resources Investigations Report 83-4238, p. 207.

param_ofpl-length-feet.aml

DEFINITION: Reports the "overland flow plane length" of each zone in the input zone map. Expressed in feet.

ORIGINAL PURPOSE: `ofp_length` (PRMS storm mode)

EXTRA DATA USED:

ArcInfo GRID of overland flow planes

SUBROUTINES CALLED:

`param_ofpl-length.aml`

DESCRIPTION:

See the description for `param_ofpl-length.aml`. The `param_ofpl-length-feet.aml` method simply runs `param_ofpl-length.aml` and then converts the units of expression to feet.

REFERENCES:

Leavesley, G.H., Lichty, R.W., Troutman, B.M., and Saindon, L.G., 1983, Precipitation-Runoff Modeling System User's Manual: U.S. Geological Survey Water-Resources Investigations Report 83-4238, p. 207.

param_one-plane_area.aml

DEFINITION: Determines size of the contributing area associated with each zone in the input zone map, expressed in the XY units of the input zone map.

ORIGINAL PURPOSE: `ca_area` (TOPNET)

EXTRA DATA USED:

SUBROUTINES CALLED:

`zone_one-plane.aml`

DESCRIPTION: (see definition)

REFERENCES: (none)

param_one-plane_ellmaj.aml

DEFINITION: Derives the contributing area associated with each zone in the input zone map, defines an ellipsoid centered on the zone, and reports the length of the major ellipsoid axis. Expressed in the XY units of the input zone map.

ORIGINAL PURPOSE: `ca_axis_maj` (TOPNET)

EXTRA DATA USED:

SUBROUTINES CALLED:

    `zone_one-plane.aml`

DESCRIPTION:

The results of this parameterization method rely on the ArcInfo Grid functions `ZONALGEOMETRY()` and `ZONALCENTROID()` for defining the ellipsoid and the axes. The reader is referred to the online documentation for ArcInfo Workstation for more information.

REFERENCES:

ESRI, Inc., 2001, ArcInfo Workstation, Version 8.1 [software].

param_one-plane_ellmin.aml

DEFINITION: Derives the contributing area associated with each zone in the input zone map, defines an ellipsoid centered on the zone, and reports the length of the minor ellipsoid axis. Expressed in the XY units of the input zone map.

ORIGINAL PURPOSE: `ca_axis_maj` (TOPNET)

EXTRA DATA USED:

SUBROUTINES CALLED:

    zone_one-plane.aml

DESCRIPTION:

The results of this parameterization method rely on the ArcInfo Grid functions `ZONALGEOMETRY()` and `ZONALCENTROID()` for defining the ellipsoid and the axes. The reader is referred to the online documentation for ArcInfo Workstation for more information.

REFERENCES:

ESRI, Inc., 2001, ArcInfo Workstation, Version 8.1 [software].

param_one-plane_ndabranch.aml

DEFINITION: Derives the contributing area associated with each zone in the user specified map of DAFLOW branches and determines which one occupies the most of each zone in the input zone map. Expressed in terms of the identification numbers of the input zone map.

ORIGINAL PURPOSE: (generic)

EXTRA DATA USED:

ArcInfo GRID of DAFLOW branches

SUBROUTINES CALLED:

`zone_one-plane.aml`

DESCRIPTION: (see DEFINITION)

REFERENCES:

Jobson, H.E., 1989, Users manual for an open-channel streamflow model based on the diffusion analogy: U.S. Geological Survey Water-Resources Investigations Report 89-4133, 73 p.

param_one-plane_perimeter.aml

DEFINITION: Derives the contributing area associated with each zone in the input zone map and reports the perimeter of this area. Expressed in the XY units of the input zone map.

ORIGINAL PURPOSE: `ca_perimeter` (TOPNET)

EXTRA DATA USED: (none)

SUBROUTINES CALLED:

```
zone_one-plane.aml
```

DESCRIPTION: (see DEFINITION)

REFERENCES: (none)

param_ov-area.aml

DEFINITION: Intersects the input zone map with a second GRID. Determines the area of each intersection. Expressed in the square XY units of the input zone map.

ORIGINAL PURPOSE: (generic)

EXTRA DATA USED:

ArcInfo GRID, set of zones to intersect with the input zone map

SUBROUTINES CALLED: (none)

DESCRIPTION:

This method intersects the input zone map and a second, user-specified GRID. The area of the intersection is reported. This is a specialty parameterization method that will produce an ASCII file without the aid of the Parameterization Engine (although it can run within the Parameterization Engine). The file will have three columns. The first is the identifier of the zone in the input zone map, the second is the identifier of the zone in the second GRID, and the third is the area of the intersection.

REFERENCES: (none)

param_ov-area-pct.aml

DEFINITION: Intersects the input zone map with a second GRID. Determines the area of each intersection and expresses this area as a percentage of the area of the zone in the input zone map.

ORIGINAL PURPOSE: (generic)

EXTRA DATA USED:

ArcInfo GRID, set of zones to intersect with the input zone map

SUBROUTINES CALLED: (none)

DESCRIPTION:

This method intersects the input zone map and a second, user-specified GRID. The area of the intersection is divided by the total area of the zone in the input zone map involved in the intersection, and is reported as a percentage.

This is a specialty parameterization method that will produce an ASCII file without the aid of the Parameterization Engine (although it can run within the Parameterization Engine). Each row will describe an intersection between the input zone map and the second GRID. Each row will have three columns. The first is the identifier of the zone in the input zone map, the second is the identifier of the zones in the second GRID, and the third is the percentage of the input map zone area represented by the intersection.

REFERENCES: (none)

param_ov-area-pct2.aml

DEFINITION: Intersects the input zone map with a second GRID. Determines the area of each intersection and expresses this area as a percentage of the area of the zone in the input zone map, and as a percentage of the area of the zone in the second GRID.

ORIGINAL PURPOSE: (generic)

EXTRA DATA USED:

ArcInfo GRID, set of zones to intersect with the input zone map

SUBROUTINES CALLED: (none)

DESCRIPTION:

This method intersects the input zone map and a second, user-specified GRID. The area of the intersection is divided by the total area of the zone in the input zone map involved in the intersection, and is reported as a percentage. The area of the intersection also is divided by the total area of the zone from the second GRID involved in the intersection to yield an additional figure.

This is a specialty parameterization method that will produce an ASCII file without the aid of the Parameterization Engine (although it can run within the Parameterization Engine). Each row will describe an intersection between the input zone map and the second GRID. Each row will have five columns. The first is and identifier for the new zone formed by the intersection, the second is the identifier of the zone in the input zone map, and the third is the identifier of the zones in the second GRID. The fourth is the percentage of the input map zone area represented by the intersection. The fifth is the percentage of the zone in the second GRID represented by the intersection.

REFERENCES: (none)

param_perimeter.aml

DEFINITION: Reports the length of the perimeter of each zone in the input zone map. Expressed in the XY units of the input zone map.

ORIGINAL PURPOSE: (generic)

EXTRA DATA USED: (none)

SUBROUTINES CALLED: (none)

DESCRIPTION: (see DEFINITION)

REFERENCES: (none)

param_poly2point.aml

DEFINITION: Assign the identity of one point-type feature to each zone in the input zone map.

ORIGINAL PURPOSE: `hru_psta, hru_tsta, basin_tsta` (PRMS); `sc_evsta, sc_psta, sc_qobsta` (TOPMODEL)

EXTRA DATA USED:

ArcInfo GRID or COVER of point type locations

SUBROUTINES CALLED:

```
data_assign-mono.aml
```

```
dimension_map_generator.aml
```

DESCRIPTION:

This method allows the user to associate a single feature from a point coverage or a GRID with each zone in the input zone map. The point-type features may be derived from a pre-existing COVER or GRID containing point-type features or may be center points derived from a pre-existing zone map. These points typically represent the locations and identities of rain, temperature, evaporation, or stream gages. The method is usually unaware of the name of the point-type features. If this is the case, then the user is required to identify the geodataset.

REFERENCES:

Beven, K. J., Quinn, P.F., Romanowicz, R.J., Freer, James, Fisher, J.I., and Lamb, R., 1995, TOPMODEL and GRIDATB – A Users Guide to the Distribution Versions (95.02): Lancaster University, United Kingdom, Centre for Research on Environmental Systems and Statistics, Institute of Environmental and Biological Sciences, 31 p.
Leavesley, G.H., Lichty, R.W., Troutman, B.M., and Saindon, L.G., 1983, Precipitation-Runoff Modeling System - User's Manual: U.S. Geological Survey Water-Resources Investigations 83-4238, 207 p.

param_poly2poly.aml

DEFINITION: Determine the most commonly occurring polygon-type feature of an input GRID within each zone (MRU) in a GRID of polygon-type features.

ORIGINAL PURPOSE: (generic)

EXTRA DATA USED:

ArcInfo GRID of polygon type features, termed "input layer"

SUBROUTINES CALLED:

```
dimension_map_generator.aml
```

DESCRIPTION:

This method performs a simple overlay analysis. For each zone in the input zone map, the identity of the single polygon-type feature from the input layer that occupies most of each zone is recorded. The polygon-type features of the input layer may be stored in a coverage or GRID.

REFERENCES: (none)

param_radpl-aspect.aml

DEFINITION: Determines the dominant type of radiation plane found in each zone of input zone map, and reports the aspect.

ORIGINAL PURPOSE: `radpl_aspect` (PRMS)

EXTRA DATA USED: (none)

SUBROUTINES CALLED:

    zone_radpl.aml

DESCRIPTION:

This method will derive the radiation planes associated with each zone of the input zone map using the zone_radpl.aml routine. The parameterization method will use extra information stored in the routine output INFO tables to return the aspect of the dominant radiation plane within each zone of the input zone map.

Radiation planes, created by zone_radpl.aml, are combinations of reclassified slope and reclassified aspect. Radiation planes are used in the computation of potential solar radiation in the PRMS. The most commonly occurring combination of reclassified slope and reclassified aspect in each zone of the input GRID of polygon-type features is used as the output.

REFERENCES:

Leavesley, G.H., Lichty, R.W., Troutman, B.M., and Saindon, L.G., 1983, Precipitation-Runoff Modeling System User's Manual: U.S. Geological Survey Water-Resources Investigations Report 83-4238, p. 207.

param_radpl-slope.aml

DEFINITION: Determines the dominant type of radiation plane found in each zone of input zone map, and reports slope.

ORIGINAL PURPOSE: radpl_slope (PRMS)

EXTRA DATA USED: (none)

SUBROUTINES CALLED:

    zone_radpl.aml

DESCRIPTION:

This method will derive the radiation planes associated with each zone of the input zone map using the zone_radpl.aml routine. The parameterization method will use extra information stored in the routine output INFO tables to return the aspect of the dominant radiation plane within each zone of the input zone map.

Table 9. Reclassification scheme for converting tree species into rooting depths.

Input tree species	Output rooting depth (inches)
White-red-jack pine	30
Spruce-fir	36
Longleaf-slash pine	48
Loblolly-shortleaf pine	48
Oak-pine	48
Oak-hickory	48
Oak-gum-cypress	30
Elm-ash-cottonwood	30
Maple-beech-birch	30
Aspen-birch	30
Douglas-fir	36
Hemlock-Sitka spruce	30
Ponderosa pine	36
Western white pine	48
Lodgepole pine	30
Larch	30
Fir-spruce	36
Redwood	36
Chaparral	30
Pinyon-juniper	30
Western hardwoods	36
Urban or Built-up Land	0
Dryland Cropland and Pasture	18
Irrigated Cropland and Pasture	18
Mixed Dryland/Irrigated Cropland and Pasture	18
Cropland/Grassland Mosaic	18
Cropland/Woodland Mosaic	18
Grassland	18
Shrubland	30
Mixed Shrubland/Grassland	24
Chaparral	24
Savanna	18
Broadleaf Deciduous Forest	36
Evergreen Coniferous Forest	36
Subalpine Forest	36
Mixed Forest	36
Deciduous Coniferous Forest	36
Evergreen Broadleaf Forest	36
Water Bodies	0
Herbaceous Wetland	18
Forested Wetland	30
Barren or Sparsely Vegetated	18
Wooded Tundra	24
Herbaceous Tundra	24
Bare Ground Tundra	18
Wet Tundra	18
Mixed Tundra	18
Perennial Snowfields or Glaciers	0

Radiation planes, created by zone_radpl.aml, are combinations of reclassified slope and reclassified aspect. Radiation planes are used in the computation of potential solar radiation in the PRMS. The most commonly occurring combination of reclassified slope and reclassified aspect in each zone of the input GRID of polygon-type features is output.

REFERENCES:

Leavesley, G.H., Lichty, R.W., Troutman, B.M., and Saindon, L.G., 1983, Precipitation-Runoff Modeling System User's Manual: U.S. Geological Survey Water-Resources Investigations Report 83-4238, p. 207.

param_rock-depth-mean-meters.aml

DEFINITION: Derives the mean depth between the soil surface and bedrock for each zone in the input zone map. Expressed in meters.

ORIGINAL PURPOSE: (generic)

EXTRA DATA USED: (none)

SUBROUTINES CALLED:

    sfc_rock-depth-max.aml

DESCRIPTION:

The maximum depth to bedrock is determined by sfc_rock-depth-max.aml for each cell as a function of the soils data at that location. This routine uses the standard ArcInfo GRID of soils found in the .../data_bin/soils/statsgo/muid, and the associated tables. Please see the "Advanced Topics" section for more information on the data bin contents. The param_rock-depth-median-meters.aml method then determines the mean of all these values within each zone of the input zone map, and converts each value into meters.

REFERENCES:

U.S. Department of Agriculture, 1994, State soil geographic (STATSGO) data base - data use information, (rev. ed.): Fort Worth, Texas, Natural Resources Conservation Service miscellaneous publication no. 1492, p. 212.

param_rock-depth-median.aml

DEFINITION: Derives the median depth between the soil surface and bedrock for each zone in the input zone map, expressed in inches.

ORIGINAL PURPOSE: (generic)

EXTRA DATA USED: (none)

SUBROUTINES CALLED:

   `sfc_rock-depth-max.aml`

DESCRIPTION:

The maximum depth to bedrock is determined by `sfc_rock-depth-max.aml` for each cell as a function of the soils data at that location. This routine uses the standard ArcInfo GRID of soils found in the .../`data_bin/soils/statsgo/muid`, and the associated tables. Please see the "Advanced Topics" section for more information on the data_bin contents. The `param_rock-depth-median.aml` method then determines the median of all these values within each zone of the input zone map.

REFERENCES:

U.S. Department of Agriculture, 1994, State soil geographic (STATSGO) data base - data use information, (rev. ed.): Fort Worth, Texas, Natural Resources Conservation Service miscellaneous publication no. 1492, p. 212.

param_rock-depth-median-meters.aml

DEFINITION: Derives the median rooting depth for each zone in the input zone map, expressed in meters.

ORIGINAL PURPOSE: (generic)

EXTRA DATA USED: (none)

SUBROUTINES CALLED:

   `sfc_rock-depth-max.aml`

DESCRIPTION:

The maximum depth to bedrock is determined by `sfc_rock-depth-max.aml` for each cell as a function of the soils data at that location. This routine uses the standard ArcInfo GRID of soils found in the .../`data_bin/soils/statsgo/muid`, and the associated tables. Please see the "Advanced Topics" section for more information on the data_bin contents. `param_rock-depth-median-meters.aml` method then determines the median of all these values within each zone of the input zone map, and converts each value into meters.

REFERENCES: (none)

param_root-depth.aml

DEFINITION: Derives the mean rooting depth for each zone in the input zone map, expressed in inches.

ORIGINAL PURPOSE: (generic)

EXTRA DATA USED: (none)

SUBROUTINES CALLED:

  `sfc_root-depth.aml`

DESCRIPTION:

Rooting depth, or the soil depth to which the roots vegetation will penetrate the soil, is determined by `sfc_root-depth.aml` for each cell as a function of the vegetation species at that location. This routine uses the standard ArcInfo GRID of tree species found in the .../`data_bin/forests/lower48`.

The default reclassification scheme is shown in table 9. If this routine is run in the interactive mode, the user is presented with this scheme and the tree species GRID. The user may modify the scheme and change the default ArcInfo GRID of tree species to be reclassified.

REFERENCES:

Powell, D.S., Faulkner, J.L., Darr, D.R., Zhu, Zhiliang, and MacCleery, D.W., 1993, Forest resources of the United States, 1992, General Technical Report RM-234 (revised): Fort Collins, Colorado, U.S. Department of Agriculture, Forest Service, 132 p.

param_root-depth-mean-meters.aml

DEFINITION: Derives the mean rooting depth for each zone in the input zone map, expressed in meters.

ORIGINAL PURPOSE: (generic)

EXTRA DATA USED: (none)

SUBROUTINES CALLED:

    `sfc_root-depth-meters.aml`

DESCRIPTION:

Rooting depth, or the soil depth to which the roots vegetation will penetrate the soil, is determined by `sfc_root-depth-meters.aml` for each cell as a function of the tree species at that location. This routine uses the standard ArcInfo GRID of tree species found in the .../`data_bin/forests/lower48`.

Subroutine `sfc_root-depth-meters.aml` runs `sfc_root-depth.aml`, which applies the default reclassification scheme, uses the depths shown in table 9 and then converts depth values to meters. The `param_root-depth-mean-meters.aml` method then determines the mean value for root depth within in each zone of the input zone map.

REFERENCES: (none)

param_root-depth-median-meters.aml

DEFINITION: Derives the median rooting depth for each zone in the input zone map, expressed in meters.

ORIGINAL PURPOSE: (generic)

EXTRA DATA USED: (none)

SUBROUTINES CALLED:

    `sfc_root-depth-meters.aml`

DESCRIPTION:

Rooting depth, or the soil depth to which the roots vegetation will penetrate the soil, is determined by `sfc_root-depth-meters.aml` for each cell as a function of the tree species at that location. This routine uses the standard ArcInfo GRID of tree species found in the .../`data_bin/forests/lower48`.

The subroutine `sfc_root-depth-meters.aml` runs `sfc_root-depth.aml`, which applies the default reclassification scheme, uses the depths shown in table 9, and then converts depth values to meters. The `param_root-depth-median-meters.aml` method then determines the median value for root depth within in each zone of the input zone map.

REFERENCES: (none)

param_sinks-pct.aml

DEFINITION: Derives a GRID of the sinks within each zone of the input zone map, calculates the total per-zone area, and reports this as a percentage (0-1.00) of the total zone area.

ORIGINAL PURPOSE: (generic)

EXTRA DATA USED: (none)

SUBROUTINES CALLED:

   `sfc_sinks.aml`

   `param_ov-area-pct1.aml`

DESCRIPTION:

The `sfc_sinks.aml` subroutine will find all the sinks (depressions or relative low points) in the DEM, and record these locations in an output GRID. The GRID of the sinks will then be intersected with the input zone map using `param_ov-area-pct1.aml` to determine the proportion of each zone that is underlain by sinks. The param_sinks-pct.aml method then reports the percentage of each zone in the input zone map that is occupied by sinks.

REFERENCES: (none)

param_slope.aml

DEFINITION: Determines the mean slope of each zone in the input zone map, expressed in percent (0-1.00).

ORIGINAL PURPOSE: `hru_slope` (PRMS)

EXTRA DATA USED:

SUBROUTINES CALLED:

`sfc_slope.aml`

DESCRIPTION:

The `sfc_slope.aml` routine will determine the slope of each cell in the DEM. See the ArcInfo Workstation online help for more on this algorithm. The results are expressed in percent. The `param_slope.aml` method will determine the mean of this product for each zone in the input zone map.

REFERENCES:

Leavesley, G.H., Lichty, R.W., Troutman, B.M., and Saindon, L.G., 1983, Precipitation-Runoff Modeling System - User's Manual: U.S. Geological Survey Water-Resources Investigations 83-4238, 207 p.

param_slope-10-85.aml

DEFINITION: Determines, for each zone in the input zone map, the slope between points located at 10 percent and 85 percent of the total length along the primary drainage path of the zone. Expressed in percent (0-1.00).

ORIGINAL PURPOSE: (generic)

EXTRA DATA USED:

ArcInfo GRID of flow direction, referred to as `%.gridfdr%`

ArcInfo GRID of the drainage network

SUBROUTINES CALLED:

(none)

DESCRIPTION:

This method creates a measure of slope that attempts to describe the tilt of a zone, as opposed to averaging of many values describing local topography. This measure was originally designed in support of the development of flood-frequency analysis.

REFERENCES: (none)

param_slope-degrees-median.aml

DEFINITION: Determines the median slope of each zone in the input zone map, expressed in percent (0-1.00).

ORIGINAL PURPOSE: (generic)

EXTRA DATA USED:

SUBROUTINES CALLED:

    `sfc_slope-degrees.aml`

DESCRIPTION:

The `sfc_slope-degrees.aml` subroutine will determine the slope of each cell in the DEM. See the ArcInfo Workstation online help for more on the `SLOPE()` algorithm. The results are expressed in degrees. The `param_slope-degrees-median.aml` method will determine the median of this product for each zone in the input zone map.

REFERENCES: (none)

param_slope-flowlength.aml

DEFINITION: Determines the slope of each zone in the input zone map, based on the flow length of the zone and the range of elevation of the zone. Expressed in percent (0-1.00).

ORIGINAL PURPOSE: (generic)

EXTRA DATA USED:

SUBROUTINES CALLED:

      `param_elevation-range-meters.aml`

      `param_flowlength.aml`

DESCRIPTION:

The `param_elevation-range-meters.aml` subroutine will determine the range of elevation for each zone in the input zone map, using the DEM. See the description for this parameterization method for more information. The `param_flowlength.aml` method will determine the length for each zone in the input zone map as a function of the flow direction. See the description for this parameterization method for more information. The range in elevation is divided by the flow length for each zone. The results are reported in meters.

REFERENCES: (none)

param_slope-median.aml

DEFINITION: Determines the median slope of each zone in the input zone map, expressed in percent (0-1.00).

ORIGINAL PURPOSE: (generic)

EXTRA DATA USED:

SUBROUTINES CALLED:

sfc_slope.aml

DESCRIPTION:

The `sfc_slope.aml` subroutine will determine the slope of each cell in the DEM. See the ArcInfo Workstation online help for more on this algorithm. The results are expressed in percent. The `param_slope-median.aml` method will determine the median of this product for each zone in the input zone map.

REFERENCES: (none)

param_snowdepletion-curve.aml

DEFINITION: Determine the snow-depletion curve for each zone in the input zone map.

ORIGINAL PURPOSE: `hru_deplcrv` (PRMS)

EXTRA DATA USED:

ArcInfo GRID of elevation

SUBROUTINES CALLED:

timber_line.aml

```
param_elevation-median-feet.aml
```

DESCRIPTION:

This method will determine the snow-depletion curve for each zone in the input zone map. The input zone map is expected to represent hydrologic response units within PRMS. Zones with a median elevation above the user-defined timberline are assigned a snow-depletion curve number of 2, those features below timber line are assigned a snow-depletion curve number of 1. Timber line defaults to 11,500 feet.

REFERENCES:

Anderson, J.R., Hardy, E.E., Roach, J.T., and Witmer, R.E., 1976, A land use and land cover classification system for use with remote sensor data: U.S. Geological Survey Professional Paper 964, 28 p.

Leavesley, G.H., Lichty, R.W., Troutman, B.M., and Saindon, L.G., 1983, Precipitation-Runoff Modeling System - User's Manual: U.S. Geological Survey Water-Resources Investigations 83-4238, 207 p.

param_snow-threshold.aml

DEFINITION: Determine the minimum snow-pack water equivalent for each zone in the input zone map, expressed in cubic inches.

ORIGINAL PURPOSE: snarea_thresh (PRMS)

EXTRA DATA USED:

ArcInfo GRID of elevation

SUBROUTINES CALLED:

```
timber_line.aml
```

```
param_elevation-median-feet.aml
```

DESCRIPTION:

The minimum snow-pack water equivalent is the threshold below which snow covers less than the full area of each zone and bare ground is exposed. This changes the albedo, radiation budgets, and other characteristics of a hydrologic response units in the PRMS model. Snowmelt is computed only for snow-covered area within a zone (that is, a hydrologic response unit) during the execution of PRMS. This relation can be treated here as either linear or constant with elevation.

The minimum snow-pack water equivalent is calculated

$$SWE_{minimum} = y + (elv_{median} - elv_{datum}) * rate, \tag{4}$$

where $SWE_{minimum}$ is the minimum snow water equivalent, y is the minimum depth of snow-pack water equivalent, elv_{median} and elv_{datum} are the median elevation (in feet) of the zone and the minimum elevation from which the rate will be applied. The *rate* is the increase in the number of inches of snow-pack water equivalent required per 1000 meters increase in elevation. The default value for y is 0. The default value for elv_{datum} is the minimum elevation for the entire watershed. The default rate is 5 inches/1000 meters. The user is presented with a menu in which y, elv_{datum}, and *rate* are editable by the user. If the method is run in the noninteractive mode, then the defaults will be used.

REFERENCES:

Anderson, J.R., Hardy, E.E., Roach, J.T., and Witmer, R.E., 1976, A land use and land cover classification system for use with remote sensor data: U.S. Geological Survey Professional Paper 964, 28 p.

Leavesley, G.H., Lichty, R.W., Troutman, B.M., and Saindon, L.G., 1983, Precipitation-Runoff Modeling System - User's Manual: U.S. Geological Survey Water-Resources Investigations 83-4238, 207 p.

param_soil-awc.aml

DEFINITION: Derives the mean soil available water-holding capacity for each zone in the input zone map, expressed in inches/inch.

ORIGINAL PURPOSE: (generic)

EXTRA DATA USED: (none)

SUBROUTINES CALLED:

    sfc_soil-awc.aml

DESCRIPTION:

The `sfc_soil-awc.aml` subroutine will determine the available water-holding capacity (AWC) of the soil, expressed in inches/inch, for every cell in the input zone map using the standard data bin GRID for soils, found at .../data_bin/soils/statsgo/muid and the associated layer INFO table. AWC for a cell is determined by averaging the low and the high values for AWC for the column of soil associated with the cell. The `param_soil-awc.aml` method will then determine the mean AWC value for each zone in the input zone map.

REFERENCES:

U.S. Department of Agriculture, 1994, State soil geographic (STATSGO) data base - data use information, (rev. ed.): Fort Worth, Texas, Natural Resources Conservation Service miscellaneous publication no. 1492, p. 212.

param_soil-bulk-density.aml

DEFINITION: Derives the mean soil bulk density for each zone in the input zone map, expressed in grams per cubic centimeter.

ORIGINAL PURPOSE: (generic)

EXTRA DATA USED: (none)

SUBROUTINES CALLED:

sfc_soil-bulk-density.aml

DESCRIPTION:

The `sfc_soil-bulk-density.aml` subroutine determines the bulk density (BD) of the soil, expressed in grams per cubic centimeter, for every cell in the input zone map using the standard data bin GRID for soils, found at …/`data_bin/soils/statsgo/muid` and the associated layer INFO table. BD is defined as the total mass of soil per unit volume.

BD for a cell is determined by averaging the low and the high values for BD for the column of soil associated with the cell. The `param_soil-bulk-density.aml` method will then determine the mean BD value for each zone in the input zone map.

REFERENCES:

U.S. Department of Agriculture, 1994, State soil geographic (STATSGO) data base - data use information, (rev. ed.): Fort Worth, Texas, Natural Resources Conservation Service miscellaneous publication no. 1492, p. 212.

param_soil-depth.aml

DEFINITION: Derives the mean soil depth for each zone in the input zone map, expressed in inches.

ORIGINAL PURPOSE: (generic)

EXTRA DATA USED: (none)

SUBROUTINES CALLED:

sfc_soil-depth.aml

DESCRIPTION:

The `sfc_soil-depth.aml` subroutine will determine the depth of the soil, expressed in inches, for every cell in the input zone map using the standard data bin GRID for soils, found at .../`data_bin/soils/ statsgo/muid` and the associated layer INFO table. Depth is defined as depth to bedrock. Depth for a cell is determined by averaging the low and the high values for depth to bedrock for the column of soil associated with the cell. The `param_soil-depth.aml` method will then determine the mean depth value for each zone in the input zone map.

REFERENCES:

U.S. Department of Agriculture, 1994, State soil geographic (STATSGO) data base - data use information, (rev. ed.): Fort Worth, Texas, Natural Resources Conservation Service miscellaneous publication no. 1492, p. 212.

param_soil-field-capacity-mean.aml

DEFINITION: Derives the mean soil field capacity for each zone in the input zone map, expressed in inches/inch.

ORIGINAL PURPOSE: (generic)

EXTRA DATA USED: (none)

SUBROUTINES CALLED:

    `sfc_soil-field-capacity.aml`

DESCRIPTION:

The `sfc_soil-field-capacity.aml` subroutine will determine the field capacity (FC) of the soil, expressed in grams per cubic centimeter, for every cell in the input zone map using the standard data bin GRID for soils, found at .../`data_bin/soils/statsgo/muid` and the associated layer INFO table.

This determination is made by adding the available water-holding capacity (see the description for `param_soil-awc.aml`) to the soil wilting point (see the description for `param_soil-wilting-point-mean.aml`). The `param_soil-field-capacity-mean.aml` method will then determine the mean soil field capacity value for each zone in the input zone map, expressed in inches/inch.

REFERENCES:

U.S. Department of Agriculture, 1994, State soil geographic (STATSGO) data base - data use information, (rev. ed.): Fort Worth, Texas, Natural Resources Conservation Service miscellaneous publication no. 1492, p. 212.

param_soil-moist.aml

DEFINITION: Derives the mean soil moisture for each zone in the input zone map, expressed in inches/inch.

ORIGINAL PURPOSE: `soil_moist_max` (PRMS)

EXTRA DATA USED: (none)

SUBROUTINES CALLED:

    sfc_soil-moist.aml

DESCRIPTION:

The `sfc_soil-moist.aml` subroutine will determine the available water holding capacity (AWC) of the soil within the soil rooting depth, expressed as the depth (inches) of water per inch of soil, for every cell in the input zone map using the standard data bin GRID for soils, found at `.../data_bin/soils/statsgo/muid` and its associated layer INFO table.

Rooting depth, or the soil depth to which vegetation will penetrate the soil, for a cell is determined as a function of the vegetation species at that location. In the interactive mode, the user is presented with the scheme, shown in table 9, to translate land cover at the surface into rooting depths. The user may modify the scheme and change the default ArcInfo GRID of land use/land cover to be reclassified.

The `sfc_soil-moist.aml` subroutine will then calculate the per-cell soil moisture value by multiplying the AWC value by the rooting depth. The `param_soil-moist.aml` method will then determine the mean soil moisture value for each zone in the input zone map.

REFERENCES:

Leavesley, G.H., Lichty, R.W., Troutman, B.M., and Saindon, L.G., 1983, Precipitation-Runoff Modeling System - User's Manual: U.S. Geological Survey Water-Resources Investigations 83-4238, 207 p.

Powell, D.S., Faulkner, J.L., Darr, D.R., Zhu, Zhiliang, and MacCleery, D.W., 1993, Forest resources of the United States, 1992, General Technical Report RM-234 (revised): Fort Collins, Colorado, U.S. Department of Agriculture, Forest Service, 132 p.

U.S. Department of Agriculture, 1994, State soil geographic (STATSGO) data base - data use information, (rev. ed.): Fort Worth, Texas, Natural Resources Conservation Service miscellaneous publication no. 1492, p. 212.

param_soil-moist-meters.aml

DEFINITION: Derives the mean soil moisture for each zone in the input zone map, expressed in meters per meter.

ORIGINAL PURPOSE: `srmax` (TOPMODEL)

EXTRA DATA USED: (none)

SUBROUTINES CALLED:

  `sfc_soil-moist-meters.aml`

DESCRIPTION:

The `sfc_soil-moist-meters.aml` subroutine will run `sfc_soil-moist.aml` and convert the units to meters. See the description for `param_soil-moist.aml` for more information. The `param_soil-moist-meters.aml` method will then determine the mean soil moisture value for each zone in the input zone map.

REFERENCES:

Beven, K. J., Quinn, P.F., Romanowicz, R.J., Freer, James, Fisher, J.I., and Lamb, R., 1995, TOPMODEL and GRIDATB – A Users Guide to the Distribution Versions (95.02): Lancaster University, United Kingdom, Centre for Research on Environmental Systems and Statistics, Institute of Environmental and Biological Sciences, 31 p.

param_soil-organic-matter.aml

DEFINITION: Derives the mean percentage of the soil content that is organic matter for each zone in the input zone map.

ORIGINAL PURPOSE: (generic)

EXTRA DATA USED: (none)

SUBROUTINES CALLED:

  `sfc_soil-organic-matter.aml`

DESCRIPTION:

The `sfc_soil-organic-matter.aml` subroutine will determine the percentage of organic matter (OM) of the soil for every cell in the input zone map using the standard data bin GRID for soils, found at .../`data_bin/soils/statsgo/muid` and the associated layer INFO table. OM for a cell is determined by averaging the low and the high values for OM for the column of soil associated with the cell. The `param_soil-organic-matter.aml` method will then determine the mean OM value for each zone in the input zone map, expressed in percent.

REFERENCES:

U.S. Department of Agriculture, 1994, State soil geographic (STATSGO) data base - data use information, (rev. ed.): Fort Worth, Texas, Natural Resources Conservation Service miscellaneous publication no. 1492, p. 212.

param_soil-pct_clay-mean.aml

DEFINITION: Determines the mean percentage of clay content in the soil associated with each zone in the input zone map.

ORIGINAL PURPOSE: (generic)

EXTRA DATA USED: (none)

SUBROUTINES CALLED:

    `sfc_soil-percent-clay.aml`

DESCRIPTION:

The `sfc_soil-percent-clay.aml` subroutine will read the value for the percentage of clay of the soil directly from the standard soils GRID in the data bin, .../`data_bin/soils/statsgo/muid`, and the associated INFO tables, for every cell. The `param_soil-pct_clay-mean.aml` method will then determine the mean value for each zone in the input zone map, expressed in percent.

REFERENCES:

U.S. Department of Agriculture, 1994, State soil geographic (STATSGO) data base - data use information, (rev. ed.): Fort Worth, Texas, Natural Resources Conservation Service miscellaneous publication no. 1492, p. 212.

param_soil-pct_sand-mean.aml

DEFINITION: Determines the average percentage of sand content in the soil associated with each zone in the input zone map, expressed in percent.

ORIGINAL PURPOSE: (generic)

EXTRA DATA USED: (none)

SUBROUTINES CALLED:

   `sfc_soil-percent-sand.aml`

DESCRIPTION:

The `sfc_soil-percent-sand.aml` subroutine will read the value for the percentage of sand of the soil directly from the standard soils GRID in the data bin, `…/data_bin/soils/statsgo/muid`, and the associated INFO tables, for every cell. The `param_soil-pct_sand-mean.aml` method will then determine the mean value for each zone in the input zone map, expressed in percent.

REFERENCES:

U.S. Department of Agriculture, 1994, State soil geographic (STATSGO) data base - data use information, (rev. ed.): Fort Worth, Texas, Natural Resources Conservation Service miscellaneous publication no. 1492, p. 212.

param_soil-pct_silt-mean.aml

DEFINITION: Determines the mean percentage of silt content associated with each zone in the input zone map.

ORIGINAL PURPOSE: (generic)

EXTRA DATA USED: (none)

SUBROUTINES CALLED:

   `sfc_soil-percent-silt.aml`

DESCRIPTION:

The `sfc_soil-percent-silt.aml` subroutine will read the value for the percentage of silt of the soil directly from the standard soils GRID in the data bin, `…/data_bin/soils/statsgo/muid`, and the associated

INFO tables, for every cell. The `param_soil-pct_silt-mean.aml` method will then determine the mean value for each zone in the input zone map, expressed in percent.

REFERENCES:

U.S. Department of Agriculture, 1994, State soil geographic (STATSGO) data base - data use information, (rev. ed.): Fort Worth, Texas, Natural Resources Conservation Service miscellaneous publication no. 1492, p. 212.

param_soil-perm.aml

DEFINITION: Determines the average soil permeability associated with each zone in the input zone map, expressed in inches of water that will infiltrate per hour.

ORIGINAL PURPOSE: (generic)

EXTRA DATA USED: (none)

SUBROUTINES CALLED:

   `sfc_soil-perm.aml`

DESCRIPTION:

The `sfc_soil-perm.aml` subroutine will determine the permeability of the soil, expressed in inches/hour, for every cell in the input zone map using the standard data bin GRID for soils, found at .../`data_bin/soils/statsgo/muid`, and the associated layer INFO table. Permeability for a cell is determined by averaging the low and the high values for permeability for the column of soil associated with the cell. The `param_soil-perm.aml` method will then determine the mean soil permeability value for each zone in the input zone map.

REFERENCES:

U.S. Department of Agriculture, 1994, State soil geographic (STATSGO) data base - data use information, (rev. ed.): Fort Worth, Texas, Natural Resources Conservation Service miscellaneous publication no. 1492, p. 212.

param_soil-perm-mean-meters.aml

DEFINITION: Determines the mean soil permeability associated with each zone in the input zone map, expressed in meters of water that will infiltrate per hour.

ORIGINAL PURPOSE: (generic)

EXTRA DATA USED: (none)

SUBROUTINES CALLED:

  `sfc_soil-perm.aml`

DESCRIPTION:

This method will run `sfc_soil-perm.aml` and convert the units of the output to meters. See the description for `param_soil-perm.aml` for more information. The `param_soil-perm-mean-meters.aml` method will then determine the mean value for each zone in the input zone map.

REFERENCES: (none)

param_soil-porosity-mean.aml

DEFINITION: Determines the mean soil porosity associated with each zone in the input zone map, expressed a value between 0 and 1.0.

ORIGINAL PURPOSE: (generic)

EXTRA DATA USED: (none)

SUBROUTINES CALLED:

  `sfc_soil-porosity.aml`

DESCRIPTION:

The `sfc_soil-porosity.aml` subroutine calculates porosity on a per-cell basis using

$$Por = 1 - (BD/2.65), \tag{5}$$

where *BD* is bulk density (see the description on `param_soil-bulk-density.aml` for more information) and *2.65* represents the assumed particle density (in grams per cubic centimeter). The `param_soil-porosity-mean.aml` method will then determine the mean porosity value for each zone in the input zone map.

REFERENCES:

U.S. Department of Agriculture, 1994, State soil geographic (STATSGO) data base - data use information, (rev. ed.): Fort Worth, Texas, Natural Resources Conservation Service miscellaneous publication no. 1492, p. 212.

param_soil-rechr.aml

DEFINITION: Derives the mean soil moisture recharge capacity for the total soil depth or the first 18 inches of soil, whichever is least for each zone in the input zone map, expressed in inches/inch.

ORIGINAL PURPOSE: `soil_rechr_max` (PRMS)

EXTRA DATA USED: (none)

SUBROUTINES CALLED:

`sfc_soil-rechr.aml`

DESCRIPTION:

This parameterization method is usually used in conjunction with `param_soil-moist.aml` to generate parameters for PRMS. Please read the description for that method prior to reading this one. The `sfc_soil-rechr.aml` subroutine will multiply the per-cell available water-holding capacity by the "recharge depth," instead of the rooting depth used in `param_soil-moist.aml`. The recharge depth is calculated on a per-cell basis as the lesser of the rooting depth or 18 inches. The `param_soil-rechr.aml` method will then determine the mean value for each zone in the input zone map.

REFERENCES:

Leavesley, G.H., Lichty, R.W., Troutman, B.M., and Saindon, L.G., 1983, Precipitation-Runoff Modeling System - User's Manual: U.S. Geological Survey Water-Resources Investigations 83-4238, 207 p.
Powell, D.S., Faulkner, J.L., Darr, D.R., Zhu, Zhiliang, and MacCleery, D.W., 1993, Forest resources of the United States, 1992, General Technical Report RM-234 (revised): Fort Collins, Colorado, U.S. Department of Agriculture, Forest Service, 132 p.
U.S. Department of Agriculture, 1994, State soil geographic (STATSGO) data base - data use information, (rev. ed.): Fort Worth, Texas, Natural Resources Conservation Service miscellaneous publication no. 1492, p. 212.

param_soil-szm.aml

DEFINITION: Derives the mean "exponential storage capacity" for each zone in the input zone map, expressed in meters (representing a depth).

ORIGINAL PURPOSE: `szm` (TOPMODEL)

EXTRA DATA USED: (none)

SUBROUTINES CALLED:

    sfc_soil-szm.aml

DESCRIPTION:

The `sfc_soil-szm.aml` subroutine calculates per-cell estimates of this storage capacity by dividing the *readily drained soil porosity* by the *rate of decrease of soil transmissivity with depth*, as determined from the standard soils data in the data bin (.../`data_bin/soils/statsgo/muid` and related INFO tables).

The rate of decrease of soil transmissivity with depth can be thought of as the decrease in hydraulic conductivity with depth, due to reduced porosity. In this method, developed in collaboration with David Wolock of the U.S. Geological Survey, the value is assumed to be 4.6. The reader is referred to the source code of the `sfc_soil-szm.aml` for more information on assumptions about this value.

Readily drained soil porosity is defined as the porosity minus the field capacity. The `sfc_soil-szm.aml` subroutine runs `sfc_soil-ne.aml` (which runs `sfc_soil-porosity.aml` and `sfc_soil-field-capacity.aml`) to derive this quantity. The `sfc_soil-ne.aml` subroutine ensures that values in the output GRID have a minimum of 0.05. Please also see the method descriptions of `param_soil-porosity-mean.aml` and `param_soil-field-capacity-mean.aml` for more information.

The `param_soil-szm.aml` method will then determine the mean value for each zone in the input zone map.

REFERENCES:

Beven, K. J., Quinn, P.F., Romanowicz, R.J., Freer, James, Fisher, J.I., and Lamb, R., 1995, TOPMODEL and GRIDATB – A Users Guide to the Distribution Versions (95.02): Lancaster University, United Kingdom, Centre for Research on Environmental Systems and Statistics, Institute of Environmental and Biological Sciences, 31 p.

U.S. Department of Agriculture, 1994, State soil geographic (STATSGO) data base - data use information, (rev. ed.): Fort Worth, Texas, Natural Resources Conservation Service miscellaneous publication no. 1492, p. 212.

param_soil-texture-prms.aml

DEFINITION: Derives the dominant soil texture for each zone in the input zone map, expressed as (1 = sand; 2 = silt; 3 = clay).

ORIGINAL PURPOSE: `soil_texture` (PRMS)

DATA LAYERS USED:

SUBROUTINES CALLED:

sfc_soil-texture-prms.aml

DESCRIPTION:

The `sfc_soil-texture-prms.aml` subroutine calculates per-cell texture based on the particle size information available in the standard soils data in the data bin (.../`data_bin/soils/statsgo/muid` and its INFO tables). Cells with more than 40 percent clay are typed as clay. Those remaining soils with more than 50 percent sand are typed as sand. The remaining soils are typed silt. The `param_soil-texture-prms.aml` method will then determine the mean value for each zone in the input zone map.

REFERENCES:

Leavesley, G.H., Lichty, R.W., Troutman, B.M., and Saindon, L.G., 1983, Precipitation-Runoff Modeling System - User's Manual: U.S. Geological Survey Water-Resources Investigations 83-4238, 207 p.

U.S. Department of Agriculture, 1994, State soil geographic (STATSGO) data base - data use information, (rev. ed.): Fort Worth, Texas, Natural Resources Conservation Service miscellaneous publication no. 1492, p. 212.

param_soil-wilt-point.aml

DEFINITION: Derives the mean (soil) wilting point for each zone in the input zone map.

ORIGINAL PURPOSE: `soil_texture` (PRMS)

DATA LAYERS USED:

SUBROUTINES CALLED:

  `sfc_soil-wilt-point.aml`

DESCRIPTION:

The `sfc_soil-wilt-point.aml` subroutine calculates per-cell estimates of wilting point (wp) based on the particle size and other soils information available in the standard soils data in the data bin (.../`data_bin/soils/statsgo/muid` and its INFO tables). The empirical equation used is based on Rawls and Brakensiek (1983) and is based on the percentage sand, percentage clay, percentage organic matter, and the bulk density values. The form is

$$WP = 0.0854 - 0.0004 * PCT_{sand} + 0.0044 * PCT_{clay} + 0.0122 * PCT_{om} - 0.0182 * BD, \qquad (6)$$

where PCT_{sand} is the percentage of sand in the soil, PCT_{clay} is the percentage of clay in the soil, PCT_{om} is the percentage of organic matter in the soil, and BD is the bulk density of the soil. The `param_soil-wilting-point.aml` method will then determine the mean value for each zone in the input zone map.

REFERENCES:

Leavesley, G.H., Lichty, R.W., Troutman, B.M., and Saindon, L.G., 1983, Precipitation-Runoff Modeling System - User's Manual: U.S. Geological Survey Water-Resources Investigations 83-4238, 207 p.

Rawls, W.J., and Brakensiek, D.L., 1983 , Agricultural Management Effects on Soil Water Retention: Natural Resources Modeling Symposium, Pingree Park, Colorado, USA, October 16-21,1983 [Proceedings] p. 115-118.

U.S. Department of Agriculture, 1994, State soil geographic (STATSGO) data base - data use information, (rev. ed.): Fort Worth, Texas, Natural Resources Conservation Service miscellaneous publication no. 1492, p. 212.

param_stream-shreve.aml

DEFINITION: Determine the stream order of each link in a drainage network, using the Shreve methodology.

ORIGINAL PURPOSE: (generic)

EXTRA DATA USED:

ArcInfo GRID of flow direction, referred to as `%.gridfdr%`

SUBROUTINES CALLED:

`sfc_flow-direction.aml`

DESCRIPTION:

Table 10. Reclassification scheme for converting aspect into maximum temperature adjustments.

Input Aspect (degrees)	Output Maximum Temperature Adjustment
Flat (-1)	0.0
North (0)	-1.8
North-East (45)	-1.0
East (90)	0.0
South-East (135)	1.0
South (180)	1.2
South-West (225)	1.8
West (270)	0.0
North-West (315)	-1.0

The input zone map is assumed to represent a drainage network. This kind of map can be easily created using the **Tool Panel>>Automated Methods>>Drainage Network** tools. This method calculates the Shreve stream order of each link in a drainage network. Each link in the network should be uniquely identified. A link is typically identified as a portion of the drainage network that is found between two confluence points, between a headwater point (that is, the starting point of a headwater stream link) and a confluence, or a confluence and the outlet of the entire AOI.

REFERENCES: (none)

param_stream-strahler.aml

DEFINITION: Determine the stream order of each link in a drainage network, using the Strahler methodology.

ORIGINAL PURPOSE: (generic)

EXTRA DATA USED:

ArcInfo GRID of flow direction, referred to as `%.gridfdr%`

SUBROUTINES CALLED:

`sfc_flow-direction.aml`

DESCRIPTION:

The input zone map is assumed to represent a drainage network. This kind of map can be easily created using the **Tool Panel>>Automated Methods>>Drainage Network** tools. This method calculates the Strahler stream order of each link in a drainage network. Each link in the network should be uniquely identified. A link is typically identified as a portion of the drainage network that is found between two confluence points, between a headwater point (that is, the starting point of a headwater stream link) and a confluence, or a confluence and the outlet of the entire AOI.

REFERENCES:

Strahler, A. N., 1952, Dynamic basis of geomorphology: Geological Society of America Bulletin, vol. 63, no. 9, 923-938 p.

param_temp-adj-max.aml

DEFINITION: Set an adjustment for each zone in the input zone map, based on aspect, for the maximum daily temperature. Expressed in degrees Fahrenheit.

ORIGINAL PURPOSE: `tmax_adj` (PRMS)

EXTRA DATA USED: (none)

SUBROUTINES CALLED:

`sfc_temp-adj-max.aml`

DESCRIPTION:

Within the PRMS model, maximum daily temperature values that have been observed at point locations can be assumed to have occurred at additional locations within the AOI (that is, such as on hillsides). Because such hillsides may not face the same direction, or have the same *aspect*, as the location where the temperature was observed, the actual temperature of the hillsides are not likely to match the observation. Therefore, PRMS allows the temperature values to be adjusted. The `sfc_temp-adj-max.aml` subroutine will, on a per-cell basis, derive an adjustment factor based on the aspect of each cell, as shown in table 10, and will then determine the most commonly occurring adjustment value for each zone in the input zone map.

If the parameterization method is run in the interactive mode, then the user has the opportunity to change the scheme presented in table 10. Otherwise, the scheme will be applied without prompting the user.

REFERENCES:

Leavesley, G.H., Lichty, R.W., Troutman, B.M., and Saindon, L.G., 1983, Precipitation-Runoff Modeling System - User's Manual: U.S. Geological Survey Water-Resources Investigations 83-4238, 207 p.
Rosenberg, N.J. Blad, B.L., Verma, S.B., 1983, Microclimate - The Biological Environment (2nd ed.): Chicago, John Wiley & Sons, Inc., p. 170.

param_temp-adj-min.aml

DEFINITION: Set an adjustment for each zone in the input zone map, based on aspect, for the minimum daily temperature. Expressed in degrees Fahrenheit.

ORIGINAL PURPOSE: `tmin_adj` (PRMS)

EXTRA DATA USED: (none)

SUBROUTINES CALLED:

`sfc_temp-adj-min.aml`

DESCRIPTION:

Please see the description for `param_temp-max-adj.aml` prior to reading the current description. The `param_temp-min-adj.aml` method is almost always used in concert with that method. It is derived in an identical manner, and although the adjustment indicated by this parameter applies only to the value observed for the minimum daily temperature (instead of the maximum), it uses the same scheme as that shown in table 10.

REFERENCES:

Leavesley, G.H., Lichty, R.W., Troutman, B.M., and Saindon, L.G., 1983, Precipitation-Runoff Modeling System - User's Manual: U.S. Geological Survey Water-Resources Investigations 83-4238, 207 p.
Rosenberg, N.J. Blad, B.L., Verma, S.B., 1983, Microclimate - The Biological Environment (2nd ed.): Chicago, John Wiley & Sons, Inc., p. 170.

param_topmodel-ach.aml

DEFINITION: The input zone map needs to represent TOPMODEL channel-routing increments, produced with the **Tool Panel>>Automated Methods>>Topmodel-type Zones>>Channel Increments** tool. Determines the percentage of the subcatchment area surrounding each zone in the input zone map that flows into that zone.

ORIGINAL PURPOSE: ach (TOPMODEL, used in conjunction with param_topmodel-ach-d.aml)

EXTRA DATA USED: (none)

SUBROUTINES CALLED:

sfc_loni-fac.aml

param_sum.aml

DESCRIPTION:

If the input zone map was not created with the **Tool Panel>>Automated Methods>> Topmodel-type Zones>>Channel Increments** tool, then the parameterization method will fail. The returned values are read directly from the VAT associated with that GRID.

REFERENCES:

Beven, K. J., Quinn, P.F., Romanowicz, R.J., Freer, James, Fisher, J.I., and Lamb, R., 1995, TOPMODEL and GRIDATB – A Users Guide to the Distribution Versions (95.02): Lancaster University, United Kingdom, Centre for Research on Environmental Systems and Statistics, Institute of Environmental and Biological Sciences, 31 p.

param_topmodel-ach-d.aml

DEFINITION: The input zone map needs to represent TOPMODEL channel-routing increments, produced with the **Tool Panel>>Automated Methods>>Topmodel-type Zones>>Channel Increments** tool. Reports the percentage of the subcatchment area surrounding each zone in the input zone map that is above that zone *and* the distance to the outlet of the subcatchment surrounding each zone in the input zone map from that zone.

ORIGINAL PURPOSE: `ach, d` (TOPMODEL)

EXTRA DATA USED: (none)

SUBROUTINES CALLED:

`param_topmodel-ach.aml`

`param_topmodel-d.aml`

DESCRIPTION:

If the input zone map was not created with the **Tool Panel>>Automated Methods>> Topmodel-type Zones>>Channel Increments** tool, then the parameterization method will fail.

This parameterization method is a convenience for running `param_topmodel-ach.aml` and `param_topmodel-d.aml`. Please see the descriptions of those methods for more information.

This is a specialty parameterization method. It will produce ASCII file output without the aid of the Parameterization Engine (although it can run within the Parameterization Engine). It will produce a four-column file. The first is the zone identifier. The second is the identifier of the surrounding subcatchment. The third is the zone ach value (that is, percentage of the subcatchment area surrounding each zone in the input zone map that flows into that zone), the fourth is the d value (that is, distance to the outlet of the subcatchment surrounding the zone).

REFERENCES:

Beven, K. J., Quinn, P.F., Romanowicz, R.J., Freer, James, Fisher, J.I., and Lamb, R., 1995, TOPMODEL and GRIDATB – A Users Guide to the Distribution Versions (95.02): Lancaster University, United Kingdom, Centre for Research on Environmental Systems and Statistics, Institute of Environmental and Biological Sciences, 31 p.

param_topmodel-d.aml

DEFINITION: Determines the distance to the outlet of the subcatchment surrounding each zone in the input zone map from that zone. The input zone map needs to represent TOPMODEL channel-routing increments produced with the **Tool Panel>>Automated Methods>>Topmodel-type Zones>>Channel Increments** tool.

ORIGINAL PURPOSE: `d` (TOPMODEL, used with `param_topmodel-ach-d.aml`)

EXTRA DATA USED: (none)

SUBROUTINES CALLED:

`sfc_flowlength-down.aml`

DESCRIPTION:

If the input zone map was not created with the **Tool Panel>>Automated Methods>> Topmodel-type Zones>>Channel Increments** tool, then the parameterization method will fail. The returned values are read directly from the VAT associated with that GRID.

REFERENCES:

Beven, K. J., Quinn, P.F., Romanowicz, R.J., Freer, James, Fisher, J.I., and Lamb, R., 1995, TOPMODEL and GRIDATB – A Users Guide to the Distribution Versions (95.02): Lancaster University, United Kingdom, Centre for Research on Environmental Systems and Statistics, Institute of Environmental and Biological Sciences, 31 p.

param_traveltime.aml

DEFINITION: Estimates the number of days a volume of water needs to pass through each link in a drainage network.

ORIGINAL PURPOSE: (generic)

EXTRA DATA USED: (none)

SUBROUTINES CALLED:

`param_velocity.aml`

`param_flowlength.aml`

DESCRIPTION:

Travel time (t) is calculated

$$t = 0.000038 \frac{L}{V}, \tag{7}$$

where L is the length of a link in the drainage network and V is that link velocity. Refer to descriptions for `param_flowlength.aml` and `param_velocity.aml`, respectively, for more information.

The input zone map is assumed to represent a drainage network, where a unique identifier is used for each link in the network. A link is defined as the path between two confluence points, between a headwater point and a confluence, or a confluence and an outlet point. The output of this method was developed for an undocumented Muskingum routing model being applied to a mountainous region.

REFERENCES:

Linsley, R.K., Kohler, M.A., Paulhus, J.L.H., 1982, Hydrology for Engineers (3rd ed.): New York, McGraw Hill, 508 p.

param_tree-dom.aml

DEFINITION: Determines the most commonly occurring tree species for each zone in the input zone map, expressed according to the code scheme for the standard tree species GRID in the data bin.

ORIGINAL PURPOSE: (generic)

EXTRA DATA USED:

ArcInfo GRID of tree species, .../`data_bin/forests/lower48`

SUBROUTINES CALLED: (none)

DESCRIPTION:

This method will determine the most commonly occurring tree species for each zone in the input zone map. The GRID of tree species is taken from the standard data bin GRID for tree species (.../`data_bin/forests/lower48`).

REFERENCES:

Table 11. Reclassification scheme for converting stream order into Manning's n and hydraulic radius values.

Stream Order	Manning's n		
1	0.052	$\frac{0.45}{R^{2/3}}$	$R^{2/3} * \frac{12,891.49}{n}$
2–3	0.035	0.71	30.23
>3	0.029	1.31	67.31

Powell, D.S., Faulkner, J.L., Darr, D.R., Zhu, Zhiliang, and MacCleery, D.W., 1993, Forest resources of the United States, 1992, General Technical Report RM-234 (revised): Fort Collins, Colorado, U.S. Department of Agriculture, Forest Service, 132 p.

param_velocity.aml

DEFINITION: Estimates the velocity of water movement of each zone in the input zone map. The input zone map is assumed to depict the links in a drainage network.

ORIGINAL PURPOSE: (generic)

EXTRA DATA USED: (none)

SUBROUTINES CALLED:

param_line-slope.aml

param_velocity-coefficient.aml

DESCRIPTION:

Velocity (V) is calculated

$$V = \sqrt{s} * R^{2/3} * \frac{1.49}{n} \quad , \tag{8}$$

where R is the hydraulic radius and n is Manning's roughness coefficient. Variable s is the slope derived by param_line-slope.aml, which divides the range in elevation for each link in the drainage network by the range in flow length for that link.

Variables R and n are estimated by param_velocity-coefficient.aml using a lookup table based on stream order, shown in table 11. The values of n in this table are based on Linsley and others (1982). The values of R were estimated by the developers. The far right column of this table is termed the "velocity coefficient" here.

The input zone map should depict a drainage network where a unique identifier has been assigned to each link in the network. A link is defined as the path between two confluence points, between a headwater point and a confluence, or a confluence and an outlet point. The output of this method was developed for an undocumented Muskingum routing model being applied to a mountainous region.

REFERENCES:

Linsley, R.K., Kohler, M.A., Paulhus, J.L.H., 1982, Hydrology for Engineers (3rd ed.): New York, McGraw Hill, 508 p.

param_velocity-coefficient.aml

DEFINITION: Estimates a coefficient for determining the velocity of water movement of each zone in the input zone map. The input zone map is assumed to depict the links in a drainage network.

ORIGINAL PURPOSE: (generic)

EXTRA DATA USED: (none)

SUBROUTINES CALLED:

`zone_order-shreve.aml`

DESCRIPTION:

This parameterization method is not usually directly invoked by a user. It is designed as a submethod to param_velocity.aml. Please see the description for `param_velocity.aml` for more information.

The values returned by this method are a function of the Shreve stream ordering (see the description for `param_stream-shreve.aml` for more information), and can be found in the far right column of table 11.

REFERENCES:

Linsley, R.K., Kohler, M.A., Paulhus, J.L.H., 1982, Hydrology for Engineers (3rd ed.): New York, McGraw Hill, 508 p.

param_wcov-trans.aml

DEFINITION: Derives the mean winter canopy transmissivity for each zone in the input zone map, expressed as a value between 0.00 - 1.00.

ORIGINAL PURPOSE: `rad_trncf` (PRMS, replaced by `param_wcov-trans2.aml`)

EXTRA DATA USED: (none)

SUBROUTINES CALLED:

`sfc_wcov-trans.aml`

DESCRIPTION:

The sfc_wcov-trans.aml subroutine will determine winter vegetation density using sfc_cov-den-win.aml (see param_cov-den-win.aml for more information) and then use this to calculate winter canopy transmissivity ($trans_{wcov}$) as

$$trans_{wcov} = 2.718^{wd} *0.09917, \tag{9}$$

where wd is winter vegetation density. Any land surface at an elevation exceeding timberline is assigned a winter canopy transmissivity of 1.0. Timberline defaults to 11,500 feet, but can be adjusted by the user if the method is run in the interactive mode. The param_wcov-trans.aml method will then determine the mean transmissivity value between 0.00 and 1.00 for each zone in the input zone map.

REFERENCES:

Leavesley, G.H., Lichty, R.W., Troutman, B.M., and Saindon, L.G., 1983, Precipitation-Runoff Modeling System - User's Manual: U.S. Geological Survey Water-Resources Investigations 83-4238, 207 p.

Miller, D.H., 1959, Transmission of insolation through pine forest canopy as it effects the melting of snow: Mitteilungen der Schweizerischen Anstalt für das forstliche Versuchwesen, Versuchsw. Mitt., v. 35, p. 35-79.

Vézina, P.E., and Péch, G.Y., 1964, Solar radiation beneath conifer canopies in relation to crown closure: Forest Science, v. 10, no. 4, p. 443-451.

param_wcov-trans2.aml

DEFINITION: Derives the mean winter canopy transmissivity for each zone in the input zone map, expressed as values between 0.00 - 1.00.

ORIGINAL PURPOSE: rad_trncf (PRMS)

EXTRA DATA USED: (none)

SUBROUTINES CALLED:

param_wcov-trans2-density.aml

DESCRIPTION:

This parameterization method is an improvement of param_wcov-trans.aml. It uses an alternative method for deriving vegetation density and modifies this value with an additional coefficient. The param_wcov-trans2-density.aml subroutine will determine the per-cell winter vegetation density using sfc_cov-den-win.aml (see param_cov-den-win.aml for more information) and then use this to calculate winter canopy transmissivity ($trans_{wcov}$) on a per cell basis. This takes the form of

$$trans_{wcov} = 2.718^{wd*-2.7557} *0.9917, \tag{10}$$

where *wd* is winter vegetation density. Any land surface at an elevation exceeding timberline is assigned a winter canopy transmissivity of 1.0. Timberline defaults to 11,500 feet, but can be adjusted by the user if they run this method in the interactive mode. The `param_wcov-trans2.aml` method will then determine the mean transmissivity value between 0.00 and 1.00 for each zone in the input zone map.

Table 12. Reclassification scheme for converting cover type to leaf loss.

Input Cover Type	Output Leaf Keep
Bare	0
Grass	80
Shrub	70
Deciduous Trees	60
Coniferous Trees	100

www.ingramcontent.com/pod-product-compliance
Lightning Source LLC
Chambersburg PA
CBHW081442170526
45166CB00008B/2288